Introduction to Optics I

Interaction of Light with Matter

Synthesis Lectures on Materials and Optics

Introduction to Optics I: Interaction of Light with Matter
Ksenia Dolgaleva

ISBN: 978-3-031-01259-4   paperback
ISBN: 978-3-031-02387-3   ebook
ISBN: 978-3-031-00251-9   hardcover

DOI: 10.1007/978-3-031-02387-3

A Publication in the Springer series
SYNTHESIS LECTURES ON MATERIALS AND OPTICS

Lecture #7
Series ISSN
Synthesis Lectures on Materials and Optics
Print 2691-1930   Electronic 2691-1949

Introduction to Optics I

Interaction of Light with Matter

Ksenia Dolgaleva
University of Ottawa

SYNTHESIS LECTURES ON MATERIALS AND OPTICS #7

ABSTRACT

This book, *Introduction to Optics I: Interaction of Light with Matter*, is the first book in a series of four covering the introduction to optics and optical components. The author's targeted goal for this series is to provide clarity for the reader by addressing common difficulties encountered while trying to understand various optics concepts. This first book is organized and written in a way that is easy to follow, and is meant to be an excellent first book on optics, eventually leading the way for further study. Those with technical backgrounds as well as undergraduate students studying optics for the first time can benefit from this book series.

The current book includes three chapters on light and its characteristics (Chapter 1), on matter from the standpoint of optics (Chapter 2), and on the interaction of light with matter (Chapter 3). Among the characteristics of light, the ones characterizing its speed, color, and strength are covered. The polarization of light will be covered in the next book of the series, where we discuss optical components. Chapter 2 discusses various atomic and molecular transitions activated by light (optical transitions). Different kinds of natural bulk material media are described: crystalline and amorphous, atomic and molecular, conductive and insulating. Chapter 3 on the interaction of light with matter describes naturally occurring phenomena such as absorption, dispersion, and nonlinear optical interactions. The discussion is provided for the natural bulk optical materials only. The interfaces between various materials will be covered in the next book on optical components.

The following three books of the series are planned as follows. In the second book, we will focus on passive optical components such as lenses, mirrors, guided-wave, and polarization optical devices. In the third book, we will discuss laser sources and optical amplifiers. Finally, the fourth book in the series will cover optoelectronic devices, such as semiconductor light sources and detectors.

KEYWORDS

optics, light, interaction of light with matter, characteristics of light, optical transitions

To Natalia, my eldest daughter.

Contents

Preface

Dear Reader: I would like to present the series of books *Fundamentals of Optics* to introduce fundamental concepts of optics and principles of operations of optical components in a simple, accessible manner. As a high school student, and later as an undergraduate student in physics, I was fascinated by optics, and decided to pursue it as a career. However, while studying optics I found that many popular textbooks lacked clarity of presentation. I had to work hard to obtain an in-depth understanding of some of the more advanced concepts. Developing intuition and links between math equations and physical reality represented an even more significant challenge. Even a very well-written book can have certain parts that require better explanation. That is why I believe that many students, especially those who are studying optics for the first time, can benefit from a textbook that focuses on clarifying difficulties in understanding the basics of optics.

In this book series, I present a concise course of optics, starting from the fundamentals of light–matter interaction and ending with practical optical components. This book is the first in the series; it is dedicated to the interaction of light with matter. The next book in the series will consider key passive optical components, which are devices that can stir light without changing its strength (optical power). The third book of the series will describe the principles of the operation of laser sources. The final, fourth book will focus on semiconductors' optics: specifically, we will discuss the principles of operation of semiconductor light sources and detectors. In the *Fundamentals of Optics* book series, I intend to explain things differently. The series, in no way, targets presentation at a significant level of detail. Rather, it is simple and designed to grasp the big picture where different parts come together coherently in a way that makes sense. The level of difficulty and detail is appropriate for an undergraduate course in optics, or for a reader with a technical background who is learning optics for the first time.

The present book, entitled *Introduction to Optics I: Interaction of Light with Matter*, consists of three chapters. Chapter 1 explains the nature of light and covers its essential characteristics, such as speed, color, and strength. Chapter 2 provides an overview of the fundamental properties of optical media and their internal structures. The concluding Chapter 3 focuses on crucial aspects of light–matter interaction. Among those are absorption and dispersion of light in optical media and nonlinear optical interactions. The latter represents a more advanced aspect of light–matter interaction; following this part can be optional. I hope that you enjoy the read, and please do not hesitate to contact me to provide feedback or to ask questions.

Ksenia Dolgaleva
October 2020

Acknowledgments

I am endlessly grateful to my students and postdoctoral researchers for their valuable feedback while the book was in progress: Stephen Ronald Harrigan, Kaustubh Vyas, Payman Rasekh, Sina Aghili, Ehsan Mobini, Soheil Zibod, Dr. Boris Rosenstein Levin, and Dr. Daniel Humberto Garcia Espinosa. I am also thankful to Natalia Shershakova for reading and providing her input as a person with a technical background from a different field in Chapter 1.

Ksenia Dolgaleva
October 2020

Symbols

c_0	Speed of light in a vacuum [m/s]
n	Refractive index of an optical medium
c	Speed of light in a medium [m/s]
λ_0	Wavelength of light in a vacuum [m]
λ	Wavelength of light in a medium [m]
k_0	Wavenumber in a vacuum [m^{-1}]
k	Wavenumber in a medium [m^{-1}]
ν	The frequency of light [Hz]
ω	Angular frequency of light [rad/s]
T	The period of a light wave [s]
$\mathcal{E}_\mathrm{p}$	The energy of a photon [J]
h	Planck constant [J·s]
$\hbar$	Reduced Planck constant [J·s]
$\mathbf{E}$	Electric field vector
$\mathbf{H}$	Magnetic field vector
$\mathbf{k}$	The wave vector
k_x	X-component of the wave vector
k_y	Y-component of the wave vector
k_z	Z-component of the wave vector
$\mathbf{r}$	Radius-vector
x	X-component of the radius-vector
y	Y-component of the radius-vector
z	Z-component of the radius-vector
$\hat{\mathbf{x}}$	The unit vector along X-axis
$\hat{\mathbf{y}}$	The unit vector along Y-axis
$\hat{\mathbf{z}}$	The unit vector along Z-axis
t	The time variable
$\varphi(\mathbf{r})$	The phase of the electromagnetic wave
$u(\mathbf{r}, t)$	Scalar wave function
$a(\mathbf{r})$	Amplitude envelope function
$\Delta\varphi$	Phase difference
$U(\mathbf{r}, t)$	Complex scalar wave function
$U(\mathbf{r})$	The time-invariant part of the complex scalar wave function

U_0	The constant complex amplitude of an optical wave
I	Optical intensity [W/m^2]
P	Optical power [W]
A	Cross-sectional area of an optical beam [m^2]
$\mathcal{E}$	Energy per pulse [J]
Δt	Temporal pulse duration [s]
τ	Temporal pulse duration [s]
$A(t)$	Complex temporal envelope
$U(t)$	Time-dependent part of the complex scalar wave function
$V(\omega)$	Spectral envelope of an optical pulse
$\Delta t_{\mathrm{FWHM}}^{\mathrm{f}}$	Field temporal full width at half-maximum of an optical pulse [s]
$\Delta \nu_{\mathrm{FWHM}}^{\mathrm{f}}$	Field spectral full width at half-maximum of an optical pulse [Hz]
Δt_{FWHM}	Intensity temporal full width at half-maximum of an optical pulse [s]
$\Delta \nu_{\mathrm{FWHM}}$	Intensity spectral full width at half-maximum of an optical pulse [Hz]
C	The constant representing the time-bandwidth product
$\Delta \lambda_{\mathrm{FWHM}}$	Intensity spectral (wavelength) full width at half-maximum of an optical pulse [m]
ν_0	The central frequency of an optical pulse [Hz]
λ_0	Also defines the central wavelength of an optical pulse [m]
$\mathcal{E}_1$	Ground energy state of a two-level system
$\mathcal{E}_2$	Excited energy state of a two-level system
$\mathbf{p}$	Dipole moment (vector)
q	Charge [C]
$\mathbf{x}$	The vector characterizing deflection of a charge from the equilibrium position
$\mathbf{x}_0$	The amplitude of a charge's deviation from the equilibrium position
$\mathbf{P}$	Polarization (vector)
N	Atomic or molecular density [m^{-3}]
$\chi(\nu)$	Optical susceptibility
$\chi'(\nu)$	The real part of the optical susceptibility
$\chi''(\nu)$	The imaginary part of the optical susceptibility
ϵ_0	The dielectric permittivity of the vacuum [F/m]
μ_0	The magnetic permeability of the vacuum [H/m]
$\mathbf{D}$	The electric displacement vector
ϵ	The dielectric permittivity of a medium [F/m]
ϵ_{r}	The relative dielectric permittivity of a medium
$\mathbf{B}$	The magnetic induction vector
μ	The magnetic permeability of a medium [H/m]
$\mathbf{M}$	The magnetization of a medium
β	Propagation constant [m^{-1}]
α	Absorption coefficient [m^{-1}]

x	The coordinate of a harmonic oscillator [m]
γ	Damping coefficient [s^{-1}]
$F(t)$	The inducing force [H]
m	The oscillator's mass [kg]
$E(t)$	Time-varying electric field driving an optical harmonic oscillator [V/m]
e	The electron's charge [C]
E_0	Time-invariable amplitude of the wave's electric field [V/m]
p	The absolute value of the dipole moment [C·m]
P	The absolute value of the polarization [C/m^2]
γ	Also represents the damping of the atomic oscillator [Hz]
χ_0	The constant entering the equation of motion of an optical oscillator
P_0	The time-invariable part of the polarization
$\Delta\nu$	The width of an optical resonance [Hz]
Q	The parameter characterizing the strength of a resonance
a	The parameter of Sellmeier equation
b_i	The parameters of Sellmeier equation
c_i	The parameters of Sellmeier equation (the resonance wavelengths [m])
β_1	Group delay per unit length [m^{-1}s]
v_g	Group velocity [m/s]
β_2	Group velocity dispersion [s^2/m]
L	The length of a dispersive medium [m]
D	Dispersion parameter [s/m^2]
ΔT	The pulse broadening [s]
τ_0	The temporal width of a transform-limited optical pulse [s]
τ	Also denotes the temporal width of a chirped optical pulse [s]
$P(t)$	The function describing the variation of the optical power within a pulse [W]
P_0	The peak power of a pulse [W]
κ	The chirp parameter
$\overleftrightarrow{\chi}$	Optical susceptibility tensor
$\overleftrightarrow{\chi}^{(1)}$	Linear optical susceptibility tensor
$\overleftrightarrow{\chi}^{(2)}$	Second-order nonlinear optical susceptibility tensor
$\overleftrightarrow{\chi}^{(3)}$	Third-order nonlinear optical susceptibility tensor
$\chi^{(1)}$	Linear optical susceptibility
$\chi^{(2)}$	Second-order nonlinear optical susceptibility [m^2/V]
$\chi^{(3)}$	Third-order nonlinear optical susceptibility [m^3/V^2]
$P^{(1)}$	Linear polarization
$P^{(2)}$	Second-order nonlinear polarization
$P^{(3)}$	Third-order nonlinear polarization

$\Delta \mathbf{k}$	Phase mismatch vector
n_o	Ordinary refractive index
n_e	Extraordinary refractive index
Δn	Birefringence
$\tilde{n}$	The overall refractive index that includes linear and nonlinear contributions
n_2	Kerr coefficient [m^2/W]
φ^{NL}	The nonlinear phase shift
P_{peak}	The peak power [W]
ω_p	Pump frequency [rad/s]
ω_s	Signal frequency [rad/s]
ω_i	Idler frequency [rad/s]

CHAPTER 1

Light

Countless examples illustrate the crucial role that light plays in our everyday lives. Visual information that we perceive with our eyes is carried and delivered in the form of light streams. Light is one of the essential factors in the life of biological species. Moreover, with the advancement of laser technologies, light can now serve as a weapon, as a material processing tool for precision manufacturing, and as a surgical tool in medicine. In this chapter, we focus on describing the fundamental nature and characteristics of light.

1.1 WAVE-PARTICLE DUALITY

As confirmed by numerous experiments, light possesses a dual nature. On the one hand, it has a wave nature and behaves as many waves co-propagating with each other and comprising a beam of light. On the other hand, it also exhibits a *corpuscular* nature and acts like a bunch of particles (*photons*) that together make a beam of light. Under what circumstances does either of these behaviors come up? Let us consider two types of experiments that reveal the dual nature of light.

1.1.1 LIGHT AS WAVES

The set of experiments in which the wave nature of light becomes evident involves *interference* and *diffraction*, whereby the light interacts with objects comparable in size with its *wavelength* (see Section 1.2.2 for the definition). Let us assume that the light emitted by a light source propagates toward an opaque screen with a circular opening. We will consider the following two situations (see Fig. 1.1): (a) the diameter of the hole in the screen is huge compared to the wavelength of light (thousands of wavelengths); and (b) the diameter of the hole is on the same order of magnitude as the wavelength of light. In the first case, we will observe a uniform cone of light behind the screen. If we put a piece of paper behind the screen with the hole, the projection of the light cone will appear on the paper as circular and uniform (see Fig. 1.1a). In the second case, where the hole's diameter is on the order of the wavelength of light, what we will see behind the screen is an interference pattern in the form of concentric light cones. One can easily visualize this pattern by placing a piece of paper behind the screen with the hole to observe the projection of the light cone: concentric rings will appear on the paper (see Fig. 1.1b). The latter phenomenon corresponds to a combined effect of interference and diffraction of waves in the light beam. It can only occur if the hole diameter is comparable to the wavelength of light. This observation can only be understood if the wave nature of light is accepted. The waves sum up in

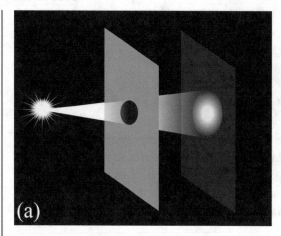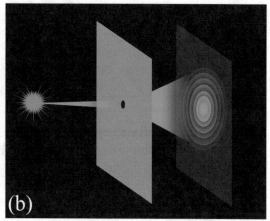

Figure 1.1: Light interaction with a hole in an opaque screen. (a) The hole diameter is thousands of wavelengths (geometrical optics limit); and (b) the hole diameter is on the order of the wavelength of light (wave optics limit).

phase in certain spatial locations, adding up to each other and producing bright rings. At other spatial locations, the waves sum up in anti-phase, thereby canceling each other and resulting in the appearance of the dark rings in the pattern. This phenomenon is better visible in the *coherent* (see Section 1.5) *monochromatic* (single-color) light. That is why the schematic in Fig. 1.1b uses a source of green light. The selection of the light source does not make any difference for the case shown in Fig. 1.1a: the outcome will be the same for either white or monochromatic light. The phenomena of interference and diffraction will be covered in the following book of the series, where we will focus on describing the interaction of light with obstacles and optical components.

1.1.2 LIGHT AS PARTICLES

Let us now consider a different experiment that illustrates the particle nature of light (see Fig. 1.2). Let us assume that the light hits a metallic surface. Assume that we have multiple light sources emitting light of different colors (quantified by *wavelengths*). In our specific example, let us assume that we have a piece of potassium as the metal. When either green or blue light hits the metal (with the wavelength values specified on the figure), a release of free electrons from the surface of the metal can occur. These electrons produce a current, which can be detected by a circuit. Also, an increase in the intensity of the incoming light does not increase the expelled photoelectrons' energy, but it increases their number. However, red light does not result in the appearance of free-electron current, even if it is much more intense than either green or blue. This phenomenon can only be understood if one accepts that the light comprises particles that carry some characteristic energies dependent on the light's color. Even if the red light is more intense with more particles in the beam of light, the electrons cannot be ejected because a

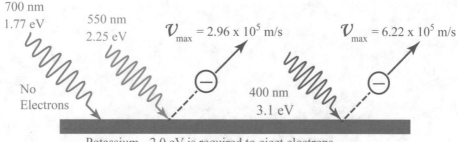

Figure 1.2: Photoelectric effect. Wavelength and energy of a photon associated with the specific color of light are shown next to the corresponding photon. v_{max} represents the maximum speed with which the electrons escape from the surface of the metal, which corresponds to the shortest-wavelength incident light. The figure is recreated based on a similar figure from the source https://www.ck12.org/physics/photoelectric-effect-in-physics/lesson/Photoelectric-Effect-CHEM/.

red-light particle has insufficient energy to expel an electron from the metal's surface. The particles of green and blue light, on the other hand, have energies sufficient to expel electrons from the metal's surface. The particles of light are called *photons*, and the effect whereby light releases free electrons from a metallic surface is called the *photoelectric effect*. One can read more on this phenomenon from the sources [1–3], which contain the effect descriptions at various levels of difficulty.

There is an essential historical fact associated with the photoelectric effect: Albert Einstein won his Nobel Prize in 1921 (see https://www.nobelprize.org/prizes/physics/1921/summary/) for providing the explanation and description to this effect involving quantization of light energy that he had performed in 1905. Nevertheless, the first work on the quantization of light belongs to a German physicist Max Planck, who studied black body radiation in the early 1900s [4]. Max Planck's observation will be discussed in the third book of this series, where we will cover active optical devices.

1.1.3 RECONCILIATION

There exists an apparent contradiction: the two sets of experiments described above, indicate different facts about what light is. On the one hand, interference and diffraction can only be understood if the wave nature of light is accepted. On the other hand, the wave representation does not explain the photoelectric effect, and one has to accept the particle nature of light to understand it. The question that arises as the result of these speculations is: What is light? Is it comprised of waves or particles? And the answer to this question is *both*. Light exhibits both wave and particle natures simultaneously. Either wave or particle nature comes up explicitly

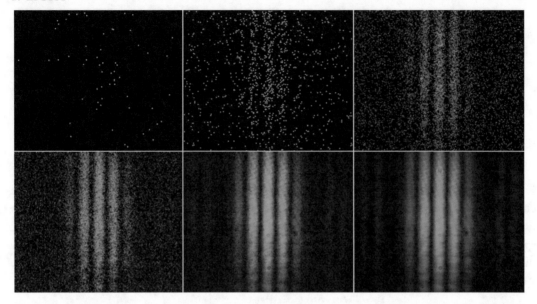

Figure 1.3: Double-slit experiment with the weak beam of light comprised of photons traveling one-at-a-time through a screen with two slits to a single-photon CCD camera. The subplots represent the image evolution as the data acquisition time becomes longer. The image is an experimental shot, courtesy of Prof. A. Weis (Université de Fribourg) and Prof. T. L. Dimitrova (Plovdiv). Source: Swiss Physical Society https://www.sps.ch/en/articles/progresses/wave-particle-duality-of-light-for-the-classroom-13/, also published in [5].

under specific experimental conditions, but they both are intrinsic to light. This property of light is termed *particle-wave duality*.

There is one experiment that can serve as reconciliation to this apparent contradiction [5]. Let us assume that a very weak beam of light interacts with a screen with two slits. The widths of the slits are identical and comparable to the wavelength of light. The beam of light is so weak that, if we accept for a moment that the light has corpuscular nature, it will correspond to photons traveling one-at-a-time and interacting with the double-slit screen individually. Imagine placing a single-photon charge-coupled device (CCD) camera, which is a camera sensitive enough to register single photons, behind the screen with the double slits (see Fig. 1.3). By allowing a sufficient time for the light to interact with the camera's sensitive elements, we will be able to observe the following time evolution of the image formed on the screen. At first, we will only be able to spot isolated spikes corresponding to single photons passing through the slits and hitting the screen (see the top-left image in Fig. 1.3), which suggests that the light beam is a stream of particles. The positions of the photon arrivals appear to be random during the initial moments of the experiment. The CCD camera can integrate images over time as the experiment goes on. After continuing the experiment for a longer time, we will observe that the photons follow a fringe pattern as they arrive at the CCD array (see the top-middle and top-right images in Fig. 1.3). After allowing even more time to pass, the fringe pattern will become very pronounced—similar to the wave formation of an interference pattern in the situation where the light is more intensive (see Fig. 1.3, bottom-right image). This observation indicates the existence of a paradox. Each photon interacts with the screen individually, as the light beam is feeble, and the photons arrive on the screen one-at-a-time. On the other hand, each photon is "aware" of interference and diffraction and traces its trajectory as if it were part of a light wave!

We can conclude from this experiment that the wave-particle duality is inherent to light, and it is valid under all circumstances, regardless of the light's intensity. The situations where the wave nature of light is prevalent correspond to higher-intensity (or higher optical power) light beams. On the other hand, the cases where the quantum nature of light becomes apparent correspond to weaker beams of light with the photons traveling one-at-a-time.

In the following section, we introduce the properties of light pertaining to its propagation speed, color, frequency, strength, and phase.

1.2   CHARACTERISTICS OF LIGHT

1.2.1   SPEED OF LIGHT

The speed at which the light travels is related to the medium where the light propagates. It is the largest in a vacuum (or in free space) where the number of molecules on the way of light is zero (or negligibly small). The value of the speed of light in a vacuum is a fundamental constant that is defined as

$$c_0 = 299{,}792{,}458 \text{ m/s}. \tag{1.1}$$

The speed of light in a medium is slowed down by a factor n characterizing the medium's refractive index. This factor exists because a material medium is comprised of molecules and atoms. The light "sees" many more obstacles on its way through the medium: interaction with these molecules and atoms takes time. To distinguish between the vacuum and non-vacuum values of the speed of light, we introduce the notation c without the subscript "0" for the speed of light in a medium. Then we have a relationship between the speed of light in the medium and vacuum:

$$c = \frac{c_0}{n}. \tag{1.2}$$

1.2.2   WAVELENGTH, FREQUENCY, AND PERIOD OF OSCILLATIONS

The fact that light has a wave nature suggests that we are dealing with oscillations. These oscillations take place both in time and space. The components of light that oscillate in time and space are its electric and magnetic fields. Light waves belong to a broader category of *electromagnetic waves*, ranging from gamma rays to radio waves. Light occupies only a narrow spectral window

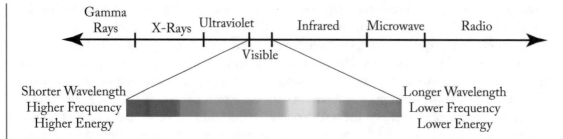

Figure 1.4: Electromagnetic wave spectrum. Gamma rays have the shortest wavelengths, highest frequencies, and highest photon energies. Radio waves lie on the opposite side of the spectrum, having the longest wavelengths, smallest frequencies, and lowest photon energies. The visible range is a narrow spectral window highlighted on the figure and expanded to show its color content. The figure is recreated based on a similar figure from the source http://gsp.humboldt. edu/olm_2015/Courses/GSP_216_Online/lesson1-2/spectrum.html.

within this range (see Fig. 1.4). Electromagnetic radiation is a form of oscillating energy, produced by charged particles, such as *electrons*, as they become accelerated by an electric field. In essence, light is electromagnetic radiation, and it consists of electromagnetic waves where the electric and magnetic fields oscillate both in time and space and carry energy. We will further refer to the electromagnetic waves associated with light as *light waves*.

Let us consider a single light wave. If we take a snapshot and assume that we can observe a wave "frozen" in time (and, consequently, in space), we will find that it exhibits a periodic change in the strength of its electric field as we move in space along the direction of the wave's propagation (see Fig. 1.5a). Looking at this snapshot, we can identify the spatial distance between the two consecutive peaks in the field oscillation to be the wavelength of light λ_0. This characteristic dictates the color that the light appears in. While the optical range typically includes a spectral window between *ultraviolet* (UV, $\lambda_0 = 300$ nm) and *infrared* (IR, $\lambda_0 = 10 \ \mu$m), defined by the operation range of conventional laser sources, the spectrum that our eye is capable of perceiving, which we call the *visible spectrum*, spans between 380 nm (purple) and 750 nm (red); see Fig. 1.6.[1]

---

[1]The latest achievements in laser sources allow us to push the optical limit to the shorter 272 nm (deep-UV) wavelengths with the new laser diode [6] and longer terahertz (THz) wavelength of around 300 μm [7]. One can also define a broader range of electromagnetic waves used in optics: from 13 nm (photolithography) to 300 μm (THz frequencies for medical diagnostics, molecular spectroscopy, airport security, and other applications).

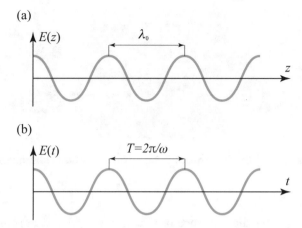

Figure 1.5: Electromagnetic light wave. (a) The variation of the electric field in space: a temporal snapshot of the wave, resulting in a wave "frozen" in space; and (b) temporal oscillations of the electric field.

Color	Wavelength
Purple	380–450 nm
Blue	450–495 nm
Green	495–570 nm
Yellow	570–590 nm
Orange	590–620 nm
Red	620–750 nm

Figure 1.6: Colors of light and their respective wavelength ranges. The figure is recreated based on a similar figure from the source http://gsp.humboldt.edu/olm_2015/Courses/GSP_216_ Online/lesson1-2/spectrum.html.

Similarly, as we did with the speed of light, we can distinguish between the wavelength in a vacuum λ_0 and the wavelength in a medium λ that is related to the wavelength in a vacuum as

$$\lambda = \frac{\lambda_0}{n}. \tag{1.3}$$

As the light enters a medium from the air, its wavelength appears to "shrink" by a factor of n (see Fig. 1.7). This happens because the speed of light in the medium is slower than its value in a vacuum. As a result, the light propagating in the vacuum starts to slow down as it enters the medium. The later wavefronts "catch up" with the earlier wavefronts that already entered the medium and experienced a slow-down. That is why the wave in the medium appears to be compressed compared to that in a vacuum, as illustrated in Fig. 1.7. An optical field (wave) is

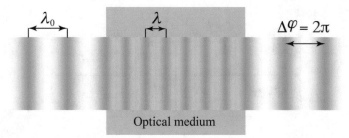

Figure 1.7: Optical wave propagation through an optical medium characterized by a refractive index $n > 1$.

shown on the figure as a periodic sequence of maxima (green) and minima (transparent); the distance between two consecutive maxima is the wavelength of the light.

By convention, the values of the wavelengths provided in association with the specific colors of light are implied to be the vacuum (or free-space) values. The colors and the corresponding ranges of wavelengths are represented in Fig. 1.6.

Another important characteristic, frequently used in optics, allows one to quantify the number of wavelengths per unit length: the *wavenumber* k_0 in a vacuum and k in the medium. The wavenumber in a vacuum is related to the wavelength of light in a vacuum as

$$k_0 = \frac{2\pi}{\lambda_0}. \tag{1.4}$$

The wavenumber in the medium is related to the wavelength of light in the medium and wavenumber in a vacuum as

$$k = \frac{2\pi}{\lambda} = \frac{2\pi n}{\lambda_0} = n k_0. \tag{1.5}$$

As the λ is shorter than λ_0, there are more wave peaks per unit distance in the medium than in a vacuum, and the wavenumber in the medium is higher than the wavenumber in a vacuum. This is explicitly illustrated in Fig. 1.7.

Another characteristic, also related to the color of light, is called the *frequency* of the oscillations ν associated with the light waves. It can be understood as follows. Let us assume that we sit at a fixed location in space and observe a light wave passing by. The wave exhibits time oscillations of its electric field. We can count how many full cycles the light wave makes in 1 sec. This will represent the value of ν. The wavelength and frequency are related through the speed of light as

$$c_0 = \lambda_0 \nu. \tag{1.6}$$

Remarkably, the frequency takes the same value both in the medium and in a vacuum: the medium does not change the frequency of electromagnetic oscillations. That is why there is no need for the subscript "0" in front of ν. The frequency counts the number of cycles per unit time; it is measured in the units of inverse seconds [s^{-1}] or hertz [Hz].

A characteristic similar to the frequency ν, the *angular frequency* of the oscillations ω, measured in radians per second or s^{-1}, is commonly used in the mathematical description of the electromagnetic wave associated with light. That is why we find it necessary to introduce it here:

$$\omega = 2\pi\nu. \tag{1.7}$$

The characteristic that describes the time duration of a single optical cycle is called the *period* of the light wave and is denoted as T (see Fig. 1.5b). The period and the frequency are inverse to each other, and they are related as

$$T = \frac{1}{\nu} = \frac{2\pi}{\omega}. \tag{1.8}$$

1.2.3   ENERGY OF A PHOTON

So far, we have been speculating about the characteristics of light pertaining to its wave nature. As we discussed earlier, the quantum nature of light co-exists with its wave nature, and, therefore, is equally important. The key characteristic associated with the quantum nature of light is the *energy of a single photon*, $\mathcal{E}_p$, defined as

$$\mathcal{E}_p = h\nu = \hbar\omega = \frac{hc_0}{\lambda_0} \tag{1.9}$$

in terms of its wavelength or frequency. The fundamental constant $h \approx 6.62607 \times 10^{-34}$ [J·s] (joules times second) is called *Planck constant*, and $\hbar = h/(2\pi) \approx 1.05457 \times 10^{-34}$ [J·s] is known as *reduced Planck constant*; they come from the quantization of electromagnetic radiation introduced by Max Planck. The energy of a photon is thus associated with the color of light: the shorter the wavelength of light, the higher the energy of the photons.

1.2.4   ELECTRIC AND MAGNETIC FIELDS

Light propagates in the form of electromagnetic waves. An electromagnetic wave consists of an *electric field* and a *magnetic field*. These fields describe the periodic oscillations of energy in space and time. The electric and magnetic field vectors, denoted by $\mathbf{E}$ and $\mathbf{H}$,[2] respectively, describe the magnitude and direction of the fields' oscillations. Electromagnetic waves are transverse: as shown in Fig. 1.8, the electric and magnetic fields oscillate in the planes perpendicular to the propagation direction. The *wave vector* $\mathbf{k}$ determines the propagation direction of the electromagnetic wave, and its magnitude $|\mathbf{k}| = k$ is simply the wavenumber, as defined in Eq. (1.5). The $\mathbf{k}$-vector has three components and can be written as $\mathbf{k} = \hat{\mathbf{x}}k_x + \hat{\mathbf{y}}k_y + \hat{\mathbf{z}}k_z$ (with $\hat{\mathbf{x}}$, $\hat{\mathbf{y}}$, and $\hat{\mathbf{z}}$ being unit vectors along X, Y, and Z axes of Cartesian coordinate frame, respectively). Each of these components describes the projection of the wave vector onto each axis.

---

[2]Through this book series, bold font represents a vector quantity, unless specified differently.

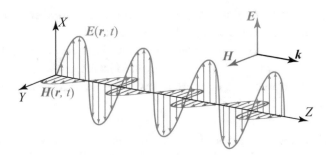

Figure 1.8: Electromagnetic wave.

The electric and magnetic field oscillations associated with the light waves can be mathematically described with a periodic function

$$\mathbf{E}(\mathbf{r}, t) = \mathbf{E}(\mathbf{r}) \cos\left[\omega t + \varphi(\mathbf{r})\right] \tag{1.10a}$$

for the electric field, and

$$\mathbf{H}(\mathbf{r}, t) = \mathbf{H}(\mathbf{r}) \cos\left[\omega t + \varphi(\mathbf{r})\right] \tag{1.10b}$$

for the magnetic field. Here, $\mathbf{r} = \hat{\mathbf{x}}x + \hat{\mathbf{y}}y + \hat{\mathbf{z}}z$ denotes the radius-vector describing the spatial coordinate variation. The functions $\mathbf{E}(\mathbf{r}, t)$ and $\mathbf{H}(\mathbf{r}, t)$ are time- and space-dependent electric and magnetic fields, respectively, and $\mathbf{E}(\mathbf{r})$ and $\mathbf{H}(\mathbf{r})$ are their corresponding *amplitude envelopes*, describing the variation of the amplitudes of the fields as a function of the coordinate. The time evolution of the electromagnetic field is fully described by the cosine function, while the function $\varphi(\mathbf{r})$ represents the phase accumulation of the electromagnetic field as it propagates; it has the form

$$\varphi(\mathbf{r}) = \mathbf{k} \cdot \mathbf{r} = k_x x + k_y y + k_z z, \tag{1.11}$$

where k_x, k_y, and k_z are the coordinates of the wave vector. Although electromagnetic waves have both electric and magnetic fields present simultaneously, it is very common to describe the properties of light, such as its *polarization*,[3] in terms of its electric field vector and strength. The strength of the electric field is measured in volts per meters [V/m].

The evolution of the electric and magnetic fields is described by two separate equations, Eqs. (1.10a) and (1.10b), respectively. It is desirable to consider these fields separately in situations where polarization matters. One such example could be the interaction of an electromagnetic wave with a crystal that exhibits *anisotropy* of its optical response, or, in other words, interacts differently with different polarizations (orientations of the electric field vector) of light.

---

[3]By polarization of light, we mean the orientation and pattern followed by the electric field vectors associated with its waves. In case if it is random from lightwave to lightwave within a light beam, the light is considered to be *unpolarized*. If it is correlated between different waves, e.g., the electric field vectors of the waves lie in the same plane, the light is said to be *polarized*. The polarization of light will be discussed in more detail in the following book of the series.

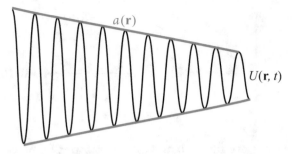

Figure 1.9: Optical field oscillations $U(\mathbf{r}, t)$ and envelope function $a(\mathbf{r})$.

Another such example is the propagation of an electromagnetic wave through an interface between two dielectric media, where we are interested in finding the fraction of the optical power reflected and refracted at the boundary. Both kinds of problems will be considered in the following book of this series.

There exist, however, situations in which the polarization of the electromagnetic field is not of importance. Among such examples are the propagation of an electromagnetic wave through a homogeneous uniform medium. In this case, it is common to introduce a *scalar wave function* $u(\mathbf{r}, t)$, used to replace Eqs. (1.10) with a single equation of the form

$$u(\mathbf{r}, t) = a(\mathbf{r}) \cos [\omega t + \varphi(\mathbf{r})] \tag{1.12}$$

that describes both the electric and magnetic field vectors. The amplitude envelope function $a(\mathbf{r})$ describes the variation of the amplitude of the electromagnetic wave as a function of the coordinates in space (see Fig. 1.9). Under the idealistic conditions of lossless propagation of a plane electromagnetic wave, it is a constant. It becomes variable in the situations where the wave experiences attenuation (its amplitude decreases with the distance, as shown in Fig. 1.9), or where the wave has a specific spatial profile, such as, e.g., *Gaussian beams*, which are optical beams emitted by lasers.

For a wave propagating along the Z-axis in a Cartesian coordinate system, the phase can simply be represented as $\varphi(\mathbf{r}) = kz$ because the wave vector does not have any transverse components ($k_x = k_y = 0$). For any wave, one can define "surfaces of equal phase" or "wavefronts"— locations along the wave separated by the phase difference of 2π. This is illustrated in Fig. 1.7 where the positions of the amplitude maxima are chosen as the surfaces of equal phase, and the phase difference between two consecutive peaks is $\Delta\varphi = 2\pi$, as marked on the figure. The surfaces of equal phase do not necessarily correspond to the maxima of the wave's field. They can be at any locations along the wave, as long as their phases differ by an integer multiple of 2π. Such locations have effectively equal phases, which reflects in the equality of the electric field at the surfaces of equal phase under the condition where the amplitude does not experience changes as the wave propagates.

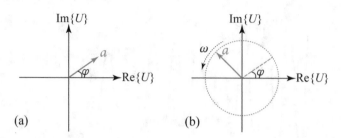

Figure 1.10: Complex scalar wave function at a fixed position **r**: (a) at a fixed moment of time t_0 (U is a fixed phasor); and (b) the time evolution of the wave function is shown as the rotation of a phasor that has an amplitude a with the angular frequency ω. The figure is recreated in similarity to Fig. 2.2-1 from *Fundamentals of Photonics* by Saleh and Teich [8].

In many books on optics, the authors also introduce a complex wavefunction $U(\mathbf{r}, t)$, given as

$$U(\mathbf{r}, t) = a(\mathbf{r}) \exp[i\varphi(\mathbf{r})] \exp(i\omega t), \tag{1.13}$$

that describes the wave completely and is related to the real wave function as

$$u(\mathbf{r}, t) = \mathrm{Re}\{U(\mathbf{r}, t)\}. \tag{1.14}$$

The time-invariant part $U(\mathbf{r})$ of the complex wavefunction can be factored out as

$$U(\mathbf{r}) = a(\mathbf{r}) \exp[i\varphi(\mathbf{r})], \tag{1.15}$$

and Eq. (1.13) can be rewritten as

$$U(\mathbf{r}, t) = U(\mathbf{r}) \exp(i\omega t) \tag{1.16}$$

with its spatial and temporal parts separated.

Why would one want to introduce yet another way of representing an optical wave? Wouldn't it be easier to follow the real (non-complex) wavefunction representation given by Eq. (1.12)? In essence, $\exp(ix) = \cos x + i \sin x$, where x is the argument of these functions. In fact, it is mathematically easier to operate with the exponential function than trigonometric functions such as sine and cosine. This serves the reason for choosing the representation given by Eq. (1.13).

Figure 1.10 represents $U(\mathbf{r}, t)$ graphically at a fixed position **r**. The X-axis of the diagram corresponds to $\mathrm{Re}\{U\}$, and the Y-axis is $\mathrm{Im}\{U\}$.

The complex scalar wave function is represented with a phasor that has an amplitude of a. The angle that the phasor makes with the $\mathrm{Re}\{U\}$-axis corresponds to the overall phase $\varphi(\mathbf{r}) + \omega t$ acquired by the electromagnetic field due to propagation in space $[\varphi(\mathbf{r})]$ and due to temporal field evolution (ωt). Figure 1.10a shows the phasor at a fixed time $t = 0$. Mathematically, it can be represented as $U = a \exp(i\varphi)$: the phasor makes a fixed angle φ with the $\text{Re}\{U\}$ axis, corresponding to the value of the overall phase of the field at $t = 0$ (when $\omega t = 0$). In Fig. 1.10b, we let the time vary while still looking at the fixed position $\mathbf{r}$, and display the temporal evolution of the wave function that can be described as $U(t) = a \exp(i\omega t)$. The overall phase of the electromagnetic field changes as ωt as the time passes by. As the result of this phase change, the phasor rotates with the angular frequency ω, describing a circle with the radius a on the $\text{Re}\{U\}$–$\text{Im}\{U\}$ diagram (shown with red dotted lines). In both examples, shown in Fig. 1.10, φ is merely the initial phase of the field at $t = 0$, and it has a constant value because we fix the position $\mathbf{r}$.

The complex wave function and its phasor representation are also described in the book *Fundamentals of Photonics* by Saleh and Teich [8] (see Chapter 2, Section 2.2.A). Here the author has provided her simplified interpretation to assist less advanced students with understanding this subject matter.

1.2.5   PLANE AND SPHERICAL WAVES

Let us consider two simple examples of monochromatic waves having specific wavefronts: planar and spherical. A monochromatic wave with planar wavefronts is shown in Fig. 1.11a. Such a wave has planes as the surfaces of equal phase, and it is called a *plane wave*. The spatial part of the wave function associated with a plane wave can be described as

$$U(\mathbf{r}) = U_0 \exp(-i\mathbf{k} \cdot \mathbf{r}), \tag{1.17}$$

where U_0 is the constant complex amplitude. A plane wave is an idealistic approximation that cannot exist in reality. Waves within a beam of light need to travel parallel to each other infinitely far (we call such waves *collimated*) in order for the plane wave model to precisely describe the beam. On the other hand, there are many situations where the plane-wave approximation holds without compromising the precision of the mathematical description. A typical example of a practical case where the plane-wave approximation is applicable would be a well-collimated beam of light where the waves travel nearly parallel to each other over a finite distance. Moreover, the plane-wave approximation represents mathematically simple and physically "well-behaved" type of waves, ideal for optical experiments and easy to model. It is, therefore, desirable to use the plane-wave approximation whenever possible to simplify mathematical description.

Spherical waves are characterized by the surfaces of equal phase being concentric circles centered around the point from which the wave emanates (see Fig. 1.11b). An ideal spherical wave can be represented as

$$U(\mathbf{r}) = \frac{U_0}{r} \exp(-i\mathbf{k} \cdot \mathbf{r}), \tag{1.18}$$

where U_0 is a complex constant and r is the distance from the wave's origin. Like plane waves, spherical waves represent an approximation where the waves are emitted by an infinitely small (point) source. This situation does not occur in nature, but the spherical wave model can still be used as an approximation in many practical situations. For example, a fundamental Gaussian

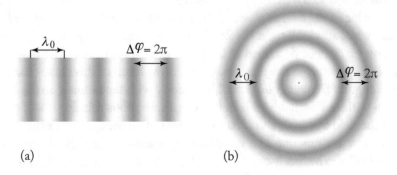

Figure 1.11: (a) Plane wave and (b) spherical wave.

beam, emitted by a laser, resembles a spherical wave at a certain distance from the beam waist, which is the narrowest place along the beam. Moreover, a sufficiently small light source can be viewed as emitting spherical waves, especially when the waves are considered far away from the source. Further, we will mostly use the complex wave function representation (where applicable).

1.2.6   OPTICAL INTENSITY, POWER, AND ENERGY

Electric and magnetic field oscillations are inherent to optical waves. As a matter of fact, there are no devices capable of directly measuring the optical field: optical oscillations have very high frequencies ($\nu \sim 10^{14}$–10^{15} Hz), while the response times of existing field meters ($\sim 10^{-6}$ s being the fastest available) are several orders of magnitude slower compared to an optical period. Since the electric field oscillates around zero, the average value measured by a field meter would also be zero. The device will not be capable of detecting real-time oscillations. In practice, it is typically the *optical power* or the *energy per pulse* associated with light (in the case where we are dealing with pulsed radiation) that is measured. These quantities are proportional to the absolute value of the electric field squared $|\mathbf{E}|^2$, and they can be measured on average, using power and energy meters. Let us discuss these characteristics pertaining to the strength and brightness of a light.

Optical intensity of an electromagnetic wave, measured in the dimensions of power per unit area (typically, in the units of W/cm^2 or W/m^2), can simply be found from its complex wave function as

$$I = |U|^2. \tag{1.19}$$

Intensity can be regarded as one of the measures of the brightness of light. At the same level of optical power, narrower optical beams (with smaller cross-sectional areas) appear to be brighter, because the same amount of power is concentrated over a smaller area.

For an optical beam with a well-defined cross-sectional area A, one can use the following approximation to relate the overall optical power P, carried by the beam, to the intensity:

$$I = \frac{P}{A}.$$
(1.20)

Most commonly, one deals in optics with Gaussian beams—the beams that have transverse electric field and intensity distributions following the Gaussian law. This suggests using a more precise relationship between the power and intensity that requires integration over the cross-sectional area:

$$P = \iint\limits_{-\infty}^{\infty} I(x,\, y)\, \mathrm{d}x\mathrm{d}y.$$
(1.21)

So far, we have been discussing continuous radiation where the light waves have infinite duration. There also exists pulsed radiation comprised of a burst of optical pulses with a finite temporal duration. In this case, one can speak of the energy carried by a single optical pulse or energy per pulse. For the situation where we have simple temporal pulse shapes (e.g., square pulses), the relationship between the optical power, the energy per pulse $\mathcal{E}$, and the temporal pulse duration Δt is simple:

$$\mathcal{E} = P\Delta t.$$
(1.22)

In some instances, an optical pulse can be approximated by a square pulse, so, Eq. (1.22) readily holds. Nevertheless, many complex pulse shapes (such as Gaussian) cannot be well approximated by a square pulse shape. In this case, an integration over the pulse duration is imperative:

$$\mathcal{E} = \int\limits_{-\infty}^{\infty} P(t)\, \mathrm{d}t.$$
(1.23)

Let us review the three characteristics, describing the strength of the optical field, but now in reverse order. The energy stored in the electromagnetic field is measured in joules [J]. It is well defined for pulsed radiation, and it is infinite for monochromatic light that is, by definition, infinite in time duration. The optical power carried by a wave is its energy per unit time, [J/s]. This characteristic is well defined for both the continuous and pulsed radiation. Finally, the optical intensity is the optical power per unit area. It is measured in the units of [W/m²] or [J/(s·m²)].

1.3   MONOCHROMATIC AND POLYCHROMATIC LIGHT

So far, we used the term "monochromatic light" several times in the text to refer to single-color infinite-duration optical waves. Monochromatic light is characterized by one particular frequency or wavelength (see Fig. 1.12a) that represents a delta-function peak in the spectral domain (see Fig. 1.12b).

Figure 1.12: (a) Monochromatic light wave and (b) its spectral representation showing an isolated frequency peak that can be represented as a delta-function. $V(\omega)$ is a spectral representation of the electric field, obtained by taking a Fourier transform of its temporal counterpart $U(\mathbf{r}, t)$.

The optical field of a monochromatic wave is given by the equation

$$U(\mathbf{r}, t) = U(\mathbf{r})\, e^{i\omega t}, \tag{1.24}$$

where ω represents the frequency (color) of the wave.

The requirements of an infinitely thin spectral line and infinite time duration cannot be met in reality. That is why monochromatic light is an idealistic model of light that still can be used as an approximation in many physical situations. One example where light can be treated as monochromatic is quasi-monochromatic light emitted by a continuous-wave laser source. It has a very narrow (delta-like) spectral function, and it lasts for a relatively long time, limited by the atomic processes in the active medium of the laser. Different continuous laser sources have different *coherence lengths*, a parameter characterizing the quality of the emitted light. The larger the coherence lengths, the higher the spectral purity (narrower spectral line) of the laser radiation. Some examples of coherence length values are 10–30 cm for helium-neon gas laser-emitting red light with $\lambda_0 = 632.8$ nm and > 1 m for argon gas laser-emitting green light with $\lambda_0 = 514.5$ nm. We will discuss more related to this matter when studying laser sources.

Polychromatic light is comprised of multiple spectral components or waves of a different wavelength. It can represent a set of discrete spectral components with different colors or a continuum of colors. An example of polychromatic light comprised of multiple discrete frequency components is shown in Fig. 1.13a, and its spectral representation is shown in Fig. 1.13b. The overall optical field in this case can be described by equation

$$U(\mathbf{r}, t) = U_1(\mathbf{r})\, e^{i\omega_1 t} + U_2(\mathbf{r})\, e^{i\omega_2 t} + U_3(\mathbf{r})\, e^{i\omega_3 t}, \tag{1.25}$$

where ω_1, ω_2, and ω_3 represent the angular frequencies of the three spectral components, while U_1, U_2, and U_3 are their respective complex amplitudes.

An example of an emission spectrum with discrete components is the spectrum of an Hg (mercury) lamp, emitting several discrete spectral lines (e.g., 253.7 nm ultraviolet, 435.8 nm blue, 546.1 nm yellow) within 200–600-nm wavelength window (see Fig. 1.14).

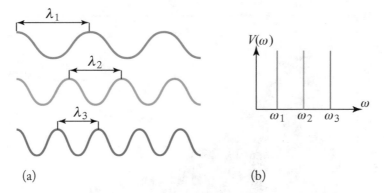

Figure 1.13: (a) Polychromatic light comprised of three different colors and (b) its spectral representation showing three frequency peaks. $V(\omega)$ is a spectral representation of the wave function, obtained by taking a Fourier transform of its temporal counterpart $U(\mathbf{r}, t)$.

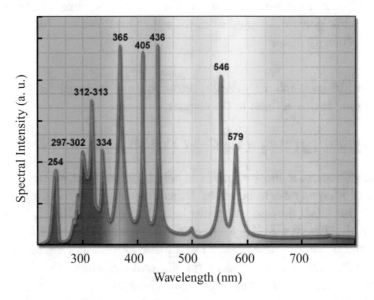

Figure 1.14: Spectrum of a mercury lamp, showing discrete spectral peaks. The figure is the property of Carl Zeiss microscopy and was reproduced from the source http://zeiss-campus. magnet.fsu.edu/articles/lightsources/mercuryarc.html with permission.

Figure 1.15 shows sunlight, which is an example of a continuous spectrum. One can see from the figure that the spectrum of the sun represents a continuum of colors rather than a set of discrete spectral components. It spans from ultraviolet to infrared. The spectral windows colored in black represent portions of the sun spectrum invisible to the human eye.

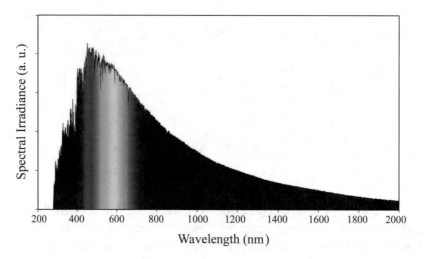

Figure 1.15: Spectrum of the sun, showing a continuum of colors. Image courtesy of Dr. Christopher S. Braid, public domain, adapted from the source http://wtamu.edu/~cbaird/sq/2013/07/03/what-is-the-color-of-the-sun/.

A finite-width (non-infinitely thin) spectrum can also be associated with pulsed radiation, which we are going to discuss later in the chapter.

1.4   WHITE LIGHT OR COLOR BLEND?

Since we already discussed monochromatic and polychromatic light, we can introduce the notion of *white light*, or what appears to be white. Different colors have specific wavelengths of light associated with them. Is it the case with white light? In fact, not. White light is a *combination* of optical waves with different wavelengths or colors. This can be easily verified by sending a beam of white light through a *dispersive element* like a prism or grating. Different colors, comprising the beam of white light, get deflected by the dispersive element at different angles, and the rainbow nature of white light can be revealed (see Fig. 1.16).

In reality, it can take less than the entire rainbow of colors to produce white light. It is sufficient to mix two specific colors of just the right shades in the right proportion. Such a color combination can be deduced from the color-mixing diagram, shown in Fig. 1.17. Mixing red, green, and blue (also called the *primary colors*) at full intensity produces white light. Cyan, magenta, and yellow are produced by combining two primary colors as follows. Cyan can be produced by mixing green and blue, magenta is a mixture of red and blue, and yellow is produced from red and green. The opposite colors, connected in Fig. 1.17 with lines, are *complementary* to each other. In order to produce white light, it is sufficient to mix only two complementary colors. It makes perfect sense, since, in each pair of complementary colors, e.g., blue and yellow,

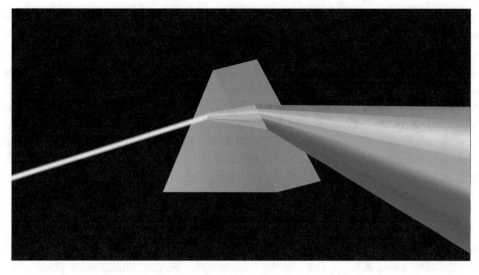

Figure 1.16: A prism separating white light into its spectral components by virtue of dispersion.

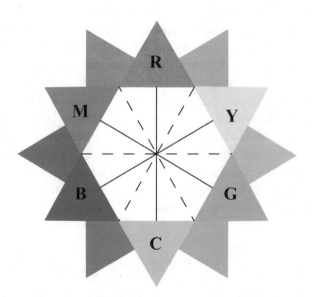

Figure 1.17: Color diagram. The primary colors are red, green and blue (schematically labeled by "R", "G", and "B" on the picture). Their respective complementary colors are cyan ("C"), magenta ("M"), and yellow ("Y"). R, G, and B mix together to produce white light. The solid lines connect primary colors with their respective complementary. The dashed lines connect mixed colors with their complementary mixed colors. The figure was reproduced in similarity to https://en.wikipedia.org/wiki/Complementary_colors.

one color is primary (blue). In contrast, the other color (yellow) is obtained by mixing the two remaining primary colors (red and green). In such a way, one effectively mixes all three primary colors, and the outcome is white light.

The color-mixing scheme we described above is termed *RGB scheme* because of the red, green, and blue adaptation as the primary colors. Remarkably, this is not the only way to represent colors. Any of the colors represented in Fig. 1.6, from purple to red, can play the role of primary colors. Any of these colors can be generated in a pure state by a light source emitting an appropriate wavelength, as specified in the table in Fig. 1.6. Which three colors to consider as the primary is a matter of choice. For example, another commonly adapted scheme relies on blue, yellow, and red as the primary colors. The white light can be represented as the three primary colors mixed, and the rest of the colors can be obtained by their various combinations. The yellow that appears as the primary, in this case, can, indeed, be the pure yellow. On the other hand, one can also achieve a visual effect of the pure yellow by mixing red and green, as with the example shown in Fig. 1.17. One can read more about color mixing and visual perception from the sources on *colorimetry*, the science used to describe the human color perception (see, for example, [9]).

The discussion we held in the previous two paragraphs is valid for the cases where we deal with light sources (with emitted light of different colors, as specified above, mixed). When it comes to colored objects (e.g., semi-transparent colored films) that do not emit light but transmit light of specific colors, the outcome of the light transmission through a combination of such objects can be completely different. As one example, overlapping three films in three primary colors and illuminating them with white (or any other color of) light will result in no light passing through them. Overlapped together, they appear to be black (see Fig. 1.18).

This effect can be understood if we take into consideration the phenomenon of *absorption* (see Section 3.1 for the detailed description). Material objects absorb light of specific colors unique to an object. If a beam of white light propagates through (or reflects off) an object that absorbs a specific color, the reflected or transmitted light will appear in the complementary color of the absorbed color. This is the reason behind the colors of objects (e.g., red apple, yellow banana). Further discussion, related to absorption and dispersion of spectral components, will follow in Chapter 3.

1.5   COHERENT AND INCOHERENT RADIATION

An important property of light waves, co-propagating with each other (originating from the same source), is their ability to form *interference patterns*. In Fig. 1.1b, we show an example of such a pattern, with a remark that it can be better observed in coherent monochromatic light. We already discussed what monochromatic light is, and in this section, we will explain coherence.

Let us take a look at Fig. 1.19 where we have three different cases of light waves co-propagating with each other and forming beams of light. In Fig. 1.19a, we deal with polychromatic light where different waves can have different wavelengths. Such light cannot exhibit

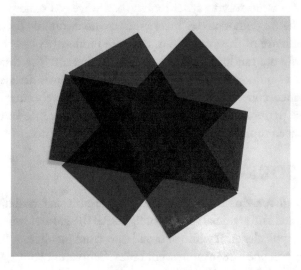

Figure 1.18: Overlap of colored films of three primary colors, red, green, and blue, results in black. No light can pass through such a combination of colored objects.

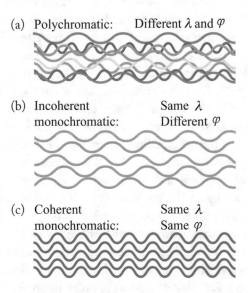

Figure 1.19: (a) Polychromatic light: the light waves in an optical beam have different wavelengths; their phases are random, which means that they are incoherent. (b) Incoherent monochromatic light, comprised of waves with the same wavelengths but random phases. (c) Coherent monochromatic light: all the light waves have the same color, and they are synchronized in phase.

coherence because there is no way to synchronize such waves to make them in phase to each other. Part (b) of the figure demonstrates incoherent monochromatic light. While all the light waves in the light beam are of the same color, they are randomly phased with respect to each other. Since their phases are random, these waves are incoherent. The third example, shown in Fig. 1.19c, demonstrates monochromatic light where all the waves are synchronized in phase. Such light can be named coherent, and such waves can effectively interfere with each other, producing interference patterns. We will talk more about interference in the next book, where we discuss more advanced topics associated with optical components.

1.6   LIGHT SOURCES

Where does coherent light originate from? Can any source emit such radiation? The answer is no. There are different natural and artificial sources of light. All the natural sources of light, including the Sun and lightning, typically emit white or broad-spectrum incoherent light. Among artificial sources, the most common ones are incandescent bulbs, fluorescent lamps, light-emitting diodes (LED), and lasers. In Fig. 1.20, we show the most common artificial sources of incoherent light.

Incandescent bulbs (see Fig. 1.20a) emit incoherent white light. These represent the first invented artificial light source powered by electricity. A sealed glass bulb with no air inside encapsulates a tungsten spiral heated by electricity and emitting light as a result of the heating. *Fluorescent light sources* [10] are another example of sources of incoherent light (see Fig. 1.20b). They are comprised of an elongated sealed glass tube with two electrodes at the opposite ends. The container is filled with different gases, depending on the desired output spectrum. The gas mixture in the bulb takes energy from the electric discharge when the bulb turns on and consumes this energy in the process of light emission. Fluorescent light sources typically emit numerous discrete spectral lines, depending on the type of gas used as the light-emitting medium. Based on the specific combination of the spectral lines, there may be different colors of the fluorescent radiation: white, blue, red, or purple (for some UV fluorescence sources).

Finally, the latest achievements in semiconductor technologies enabled robust, affordable *light-emitting diodes*, powered by an electric bias [11]. Certain kinds of semiconductors are capable of emitting light naturally when a direct electric bias is applied to them. In Fig. 1.20c, we show such a piece of semiconductor material, representing part of an LED device. While LED emission spectra are relatively broad, they are centered around some specific frequency. As a result, LEDs naturally have a specific color associated with their radiation. In Fig. 1.20d, we show the emission from LED strips of various colors. In order to produce a white-light LED [12], one has to design a light source from a combination of numerous LEDs emitting different colors that can mix to make white. Another approach toward a white-light LED is deploying *nonlinear optical interactions* (see Chapter 3, Section 3.3).

The only source of highly coherent radiation is *laser* (see Fig. 1.21) [13]. By the nature of the processes in the basis of laser operation, it emits coherent monochromatic light with a minimal divergence angle. In contrast, the sources of incoherent radiation that we discussed

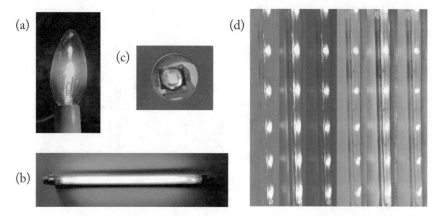

Figure 1.20: Sources of incoherent radiation: (a) incandescent light bulb; (b) fluorescent light; (c) a piece of semiconductor, representing the light-emitting element of a light emitting diode; and (d) LED strips emitting radiation of different colors.

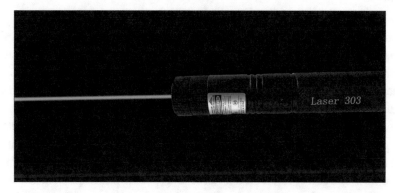

Figure 1.21: A green laser pointer emitting $\lambda_0 = 532$ nm radiation.

above emit in all directions. Even if their light has a distinct spectral content, the emitted light waves have different phases. Lasers are a subject of a significant portion of the third book of this series; we will provide a detailed description of their characteristics and principles of operation there.

1.7   CONTINUOUS AND PULSED RADIATION

Continuous radiation is regarded as light of infinite duration, which is monochromatic light. Such light is comprised of waves of a specific frequency, having delta-function-like spectra similar to the one shown in Fig. 1.12. In practice, we are dealing with quasi-monochromatic light comprised of waves of relatively long, but not infinite, duration. A laser-emitting quasi-

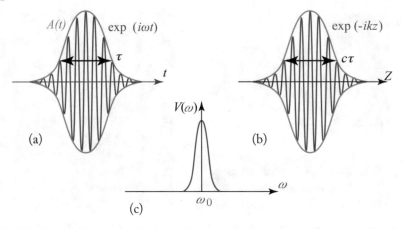

Figure 1.22: Optical pulse (a) in time domain; (b) a snapshot (as a function of z); (c) the spectrum. The figure is recreated in similarity to Fig. 2.6-2 from *Fundamentals of Photonics* by Saleh and Teich [8].

monochromatic light is said to operate in the *continuous-wave* (CW) regime. In this case, the laser output has a very narrow but not infinitely narrow spectrum. The specific spectral width is dictated by various physical mechanisms contributing to spectral broadening, which will be discussed in a later book in the course of providing the material on laser sources. However, not all light sources operate in the continuous regime. Some of them emit pulsed radiation that resembles a train of pulses of finite duration. Figure 1.22 shows an example of a simple optical pulse, a plane-wave pulse, drawn schematically as a function of time [part (a) of the figure]. It can be seen from the figure that the pulse has a finite time duration described by a time characteristic τ. We will discuss how one can define τ in relevance to a specific pulse shape below in this section.

Figure 1.22b shows the optical pulse "frozen" in space as we take a snapshot at a particular instant. Specifically, it shows the pulse extent along Z-direction, which is the propagation direction of the pulse. The pulse with the temporal duration τ has an extent $c\tau$ along the propagation direction, where c is the speed of light in the medium where the pulse propagates, as defined earlier.

The electromagnetic field associated with the optical pulse can be represented as

$$U(\mathbf{r}, t) = A(t)a(\mathbf{r}) \exp\left[i\omega_0 \left(t - \frac{z}{c}\right)\right] = U(\mathbf{r})U(t), \tag{1.26}$$

where

$$U(t) = A(t) \exp(i\omega_0 t), \tag{1.27}$$

and $U(\mathbf{r})$ is given by Eq. (1.15). Here $A(t)$ is the complex temporal envelope of the pulse (see Fig. 1.9a), describing its temporal waveform (shape), while $e^{i\omega_0 t}$ is the part associated with the optical oscillations at frequency ω_0: recall that $\exp(ix) = \cos x + i \sin x$ describes periodic oscillations. Here and below, the subscript "0" in front of the frequency ω indicates that the optical cycle associated with the radiation has the specific fixed frequency of oscillations ω_0. The term $-\omega_0 z/c$ represents the phase accumulation as the wave propagates along Z-direction. Recalling from Eq. (1.5) that $k = 2\pi/\lambda = 2\pi\nu/c = \omega/c$, we can simply write $-\omega_0 z/c = -kz$. When we are dealing with the optical wavelength range, the exponential term typically varies much more rapidly in time compared to the variation of $A(t)$, as it is evident from Fig. 1.22a. The most commonly occurring temporal waveforms $A(t)$ can be described by Gaussian, sech, and square mathematical functions. Optical pulses can also be arbitrarily shaped using special pulse-shaping techniques requiring the implementation of gratings and interferometers. It is, however, outside the scope of this book.

As we discussed above, continuous monochromatic (or quasi-monochromatic) radiation has an isolated frequency peak in the spectral domain, well described by a delta function. Due to its finite nature, pulsed radiation has a spectrum of frequencies related to the pulse's temporal characteristics, such as duration and temporal shape, via the Fourier transform:

$$V(\omega) = \int_{-\infty}^{\infty} U(t)\exp(-i\omega t)\mathrm{d}t. \tag{1.28}$$

Here $V(\omega)$ is the spectral envelope of the pulse, shown in Fig. 1.22c. The inverse relationship holds as well:

$$U(t) = \int_{-\infty}^{\infty} V(\omega)\exp(i\omega t)\mathrm{d}\omega. \tag{1.29}$$

The temporal and frequency properties of light are thus closely interrelated. An infinite-duration wave has a delta-peak frequency spectrum. On the other hand, an infinitely short (described by a temporal delta function) pulse would exhibit an infinitely broad spectrum. The shorter the pulse duration is, the broader is its spectrum, and vice versa, as long as *transform-limited* (non-chirped)[4] pulses are considered. This does not necessarily hold for *dispersion-broadened* pulses, as will be discussed later in the book. For a more advanced perspective, one can refer to the *Fundamentals of Photonics* book [8], Chapter 2, Section 2.6.

Let us discuss one of the most commonly occurring temporal pulse waveform: the Gaussian pulses. Such pulses are naturally generated in a pulsed laser system; they represent most of the cases we deal with in optics. That is why we find it instructive to discuss them in this course briefly. For a Gaussian pulse, we can assume the simplest case of a pulsed plane wave radiation, described by the wave function

$$U(\mathbf{r}, t) = U(\mathbf{r})\exp\left[-\frac{t^2}{2\tau^2}\right]\exp(i\omega_0 t). \tag{1.30}$$

---

[4]The pulse is said to be non-chirped if there is no frequency distribution across its temporal envelope so that every instant of time within the envelope contains all the frequency components of the pulse spectrum present simultaneously. Chirped optical pulses have frequency re-distribution across their temporal envelopes. For example, such pulses can have longer wavelengths (or lower frequencies) at the front of the envelope and shorter wavelengths (higher frequencies) at the back. Such pulses are said to have "red nose" and "purple tail", which are figurative expressions.

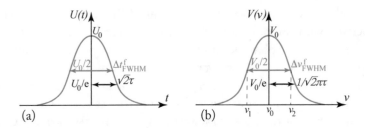

Figure 1.23: Gaussian pulse wave function (a) in time domain and (b) spectrum. $U_0 = |U(t = 0)|$ is its maximum value in time domain, $V_0 = |V(v_0)|$ is the peak in frequency domain. U_0/e (V_0/e) is the level where the exponential half-width is defined, while $U_0/2$ ($V_0/2$) is the level where FWHM is defined.

It is clear from the dependence that the portion represented by $\exp[-t^2/(2\tau^2)]$ describes the temporal characteristics of the pulse, since the complex exponential only describes the oscillations of the electromagnetic field, as discussed previously. In this particular case, the temporal pulse duration $\sqrt{2}\tau$ is the *half-width at 1/e*, which means that, if we divide the peak of the electric field by e ≈ 2.718 and take half of the width of the pulse, the corresponding temporal duration will be exactly $\sqrt{2}\tau$, as shown schematically in Fig. 1.23. However, a more practical measure of the pulse duration is its *full width at half-maximum* (FWHM). It is the full width of the pulse at one half of its maximum electric field value, and it is related to $\sqrt{2}\tau$ as

$$\Delta t_{\text{FWHM}}^{\text{f}} = 2\sqrt{2\ln 2}\,\tau. \tag{1.31}$$

In a similar manner to the characteristic pulse duration describing the temporal extent of a pulse, one can introduce *spectral width* of a pulse, describing its extent in the spectral domain. The spectral width of a pulse, coupled with its central (most powerful) frequency value (what we call v_0 in the example shown in Fig. 1.23b), carries the information about the color content of the pulse. One can perform the Fourier transform on the Gaussian field function (1.30) and verify that its spectral shape is

$$V(v) \propto \exp\left[-2\pi^2\tau^2 v^2\right], \tag{1.32}$$

which is of Gaussian shape as well! From here, one can find the exponential half-width in the spectral domain to be $1/(\sqrt{2}\pi\tau)$, and the spectral FWHM

$$\Delta v_{\text{FWHM}}^{\text{f}} = \frac{\sqrt{2\ln 2}}{\pi\tau}. \tag{1.33}$$

The beauty of Gaussian pulses is in the fact that they have a Gaussian temporal profile and a Gaussian spectrum. This makes it easy to quantify them.

So far, we have been speculating about the field pulse temporal characteristics, which is emphasized by adding the superscript "f" to the characteristics in Eqs. (1.31) and (1.33).

What we measure with devices in the laboratory is the power of the optical pulses, and the spectral width and temporal distribution of power are far more useful. According to Eq. (1.19), $I = |U(\mathbf{r}, t)|^2$, and, from Eq. (1.30), we can conclude that the optical power temporal envelope has the form $\exp(-t^2/\tau^2)$. From here, it is clear that the temporal exponential half-width of the optical pulses is τ, and the corresponding temporal FWHM is

$$\Delta t_{\text{FWHM}} = 2\sqrt{\ln 2}\,\tau. \tag{1.34}$$

One can verify that its spectral FWHM is

$$\Delta \nu_{\text{FWHM}} = \frac{\sqrt{\ln 2}}{\pi \tau}. \tag{1.35}$$

The property of the Fourier transform imposes certain limits on the minimum value of the light pulse characteristic called *time-bandwidth product* (TBP). This characteristic represents a product of the power temporal FWHM and power spectral FWHM, which must be a constant of a specific value for a given temporal waveform:

$$\Delta t_{\text{FWHM}} \Delta \nu_{\text{FWHM}} = C. \tag{1.36}$$

The constant C represents the *minimum value* this product can take, and its value is minimum for transform-limited pulses. For chirped pulses, TBP is necessarily higher than C, reflecting the longer pulse duration due to a temporal spread of the pulse. Such spread is typically caused by dispersion and nonlinearity, as will be discussed later in the book. In case of a Gaussian pulse, the constant C takes the value of 0.44, and the spectral width of a transform-limited Gaussian pulse (at full width at half-maximum) is related to its temporal FWHM duration as

$$\Delta t_{\text{FWHM}} \Delta \nu_{\text{FWHM}} \approx 0.44. \tag{1.37}$$

One can easily verify it by taking Eqs. (1.34) and (1.35) and computing the product $\Delta t_{\text{FWHM}} \Delta \nu_{\text{FWHM}}$.

The spectral width of an optical pulse can also be defined in wavelength domain. The corresponding characteristic, the wavelength full-width at half maximum ($\Delta \lambda_{\text{FWHM}}$), is related to its frequency counterpart $\Delta \nu_{\text{FWHM}}$ according to

$$\Delta \lambda_{\text{FWHM}} = \frac{c_0 |\Delta \nu_{\text{FWHM}}|}{\nu_0^2}. \tag{1.38}$$

This relationship can be verified if one admits that $\Delta \lambda_{\text{FWHM}} = \lambda_1 - \lambda_2$, where λ_1 and λ_2 correspond to the frequencies ν_1 and ν_2, taken at the spectral locations which are equidistant from the central frequency ν_0 at half-maximum: $\nu_1 = \nu_0 - \Delta \nu_{\text{FWHM}}/2$; $\nu_2 = \nu_0 + \Delta \nu_{\text{FWHM}}/2$ (see Fig. 1.23b). From Eq. (1.6), we have: $\lambda_{0,1,2} = c_0/\nu_{0,1,2}$. In the above definition of $\Delta \lambda_{\text{FWHM}}$, we subtract the shorter wavelength λ_2 from the longer wavelength λ_1 to obtain a positive

$\Delta\lambda_{\text{FWHM}}$. On the other hand, for the frequency, we have $\Delta\nu_{\text{FWHM}} = \nu_2 - \nu_1$, subtracting the lower frequency from the higher one. Most commonly, we can use the assumption that $\nu_0 \gg |\nu_0 - \nu_{1,2}|$. This allows us to use ν_0^2 in place of the product $\nu_1\nu_2$ in the denominator in Eq. (1.38). The opposite relationship for expressing $\Delta\nu_{\text{FWHM}}$ in terms of $\Delta\lambda_{\text{FWHM}}$ holds as well:

$$\Delta\nu_{\text{FWHM}} = \frac{c_0 |\Delta\lambda_{\text{FWHM}}|}{\lambda_0^2}. \tag{1.39}$$

The material provided in Section 1.7 is also described in more detail in the book *Fundamentals of Photonics* by Saleh and Teich [8] (see Appendix A, Section A.2). Here the author has provided her alternative interpretation to assist less advanced students with understanding this subject matter.

The brief overview, given in this chapter, summarizes the properties of light. The level of presentation adapted in this book is rather simplistic, aiming to assist first-time learners. There are plenty of other books the reader can refer to in order to enhance their knowledge [8, 14, 15]. In the following chapter, we proceed to describe matter from the standpoint of light propagation and interaction.

CHAPTER 2

Matter

In Chapter 1, we discussed the nature of light and some of its properties. We introduced the concepts of the speed of light, wavelength, and wavenumber while emphasizing that the values of these characteristics in a vacuum differ from those in a material medium by a factor of the medium's refractive index. In this chapter, we will discuss material media in more detail, which is necessary to describe how light propagates in the matter, how it interacts with the matter, and the accompanying phenomena associated with this interaction.

The first concept that we mention in relevance to light propagation through a medium is the medium's refractive index. It is a factor by which the speed of light in the medium gets "slowed down." The reason this slow down occurs is that the molecules and atoms, forming the material medium, represent obstacles in the way of light as it propagates through the medium. The light takes additional time to "talk" to the molecules and atoms, which results in the slower speed of its propagation even when there are no *resonances* involved. Below we discuss resonant phenomena that can occur when the frequency of light matches one of the resonances in the material medium where it propagates. Let us look into the internal structure of a material medium to understand how it responds to the incident light.

2.1 ATOMS AND ATOMIC TRANSITIONS

Any material from the periodic table has *atoms* as its structural elements. An atom is comprised of a *nucleus*, containing *protons* and *neutrons*, and *electrons*, orbiting the nucleus (see Fig. 2.1). Electrons are negatively charged elementary particles, carrying the charge of $-e = -1.60217662 \times 10^{-19}$ C, where C is an abbreviation of the charge units, *coulomb*. The protons are particles that carry the charge equal in magnitude and opposite in sign to that of an electron. An atom under normal conditions has an equal number of protons in the nucleus and electrons on the orbits, and is, therefore, an electrically neutral entity. Neutrons do not carry any charge; together with protons, they form atomic nuclei. We remark that such a representation of an atom is simplistic because it is restricted to a particular plane. In reality, the motion of electrons is similar to that of the planets orbiting the Sun: they are not localized in position, but rather form a cloud of probability where one is likely to find an electron, called *orbital*. However, in order to develop a basic understanding of the principles of interaction of light with matter, it is sufficient to adopt the simplistic representation of Fig. 2.1, which we will restrict ourselves to in this book series.

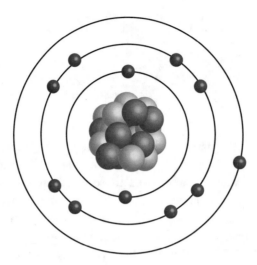

Figure 2.1: Sodium atom, containing 12 neutrons, 11 protons, and 11 electrons. The neutrons are shown with green-yellow spheres, and the protons are shown with red spheres. The two types of particles together make the nucleus. Electrons are shown with dark-blue spheres occupying the orbits around the nucleus. The image was recreated in similarity with https://www.shutterstock.com/image-illustration/3d-render-atom-structure-sodium-isolated-550082380.

Figure 2.1 shows a schematic of a sodium atom (Na in the periodic table of chemical elements) having the atomic mass of 23. It comprises 12 neutrons, 11 protons, and 11 electrons, occupying the orbits around the nucleus. The orbits are schematically shown as solid circles centered around the nucleus comprised of protons and neutrons.

Various orbits associated with an atom represent relative energies carried by the electrons residing on the orbits. The higher the energy of an electron, the higher (further from the nucleus) the orbit it occupies. An electron initially residing on a lower orbit can get excited and, as a result, can "jump up" to a higher orbit. Similarly, it can lose its excitation and can "jump down" to a lower orbit. We will further consider three transitions that could occur in atoms with light assistance. These transitions can be better understood through light's corpuscular nature, visualizing it as a collection of photons propagating together within a single optical beam.

Let us consider an atom with multiple energy levels (orbits) available for electrons to occupy. We will narrow down our scope to considering two specific energy levels, one being closer to the nucleus (the *ground state*, or the lower-energy state), and another one, representing higher energy, situated further away from the nucleus (the *excited state*). The two energy states, marked as $\mathcal{E}_1$ for the ground and $\mathcal{E}_2$ for the excited state, are schematically shown in Fig. 2.2.

Part (a) of the figure shows an atom with an electron resting in the ground state, while in Part (b) an excited atom with an electron in the excited state is displayed.

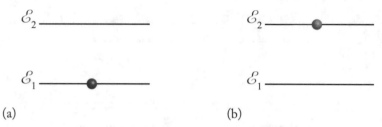

Figure 2.2: Ground ($\mathcal{E}_1$) and excited ($\mathcal{E}_2$) energy states of an atom with an electron (a) residing in the ground state and (b) being excited to the excited state.

There are three kinds of atomic transitions occurring with the assistance of light. For these transitions to occur, the photon's energy should closely match the energy difference between the excited and ground states of the atom:

$$\mathcal{E}_p = h\nu = \frac{hc}{\lambda} = \mathcal{E}_2 - \mathcal{E}_1. \tag{2.1}$$

1. **Absorption:** An atom with an electron resting initially on the ground state can absorb a photon with the energy matching the difference $\mathcal{E}_2 - \mathcal{E}_1$. After "consuming" the photon, the electron gets additional energy sufficient to make a transition from the ground state $\mathcal{E}_1$ to the excited state $\mathcal{E}_2$ because the energy of the photon satisfies the requirement of Eq. (2.1). The diagram of the process is shown in Fig. 2.3a.

2. **Spontaneous Emission:** An atom with an electron initially excited to the level $\mathcal{E}_2$ can spontaneously emit a photon with the energy satisfying Eq. (2.1). After losing this energy, the electron relaxes back to the ground state $\mathcal{E}_1$. This process is shown in Fig. 2.3b.

3. **Stimulated Emission:** An emission process can also be induced by the presence of other photons in the vicinity of an excited atom: a photon with the energy matching the difference $\mathcal{E}_2 - \mathcal{E}_1$ can "force" the electron to make a transition to the ground state while emitting a "clone" photon having the same energy, phase, polarization, and flying in the same direction as the inducing photon. As an outcome of this process, there will be two identical photons, and the electron will return to the ground state as it will lose its excitation. This process is in the basis of laser operation; we will turn to it in more detail as we discuss the relevant subject in a later book of this series. It is schematically shown in Fig. 2.3c.

As discussed above, the three atomic transitions that occur with photons' involvement have a restriction for the photons' energy to satisfy the relationship (2.1). This relationship represents the *resonance* between two specific atomic levels. Speaking of atoms, we refer to the resonances associated with their electron transitions. A different kind of resonance exists where

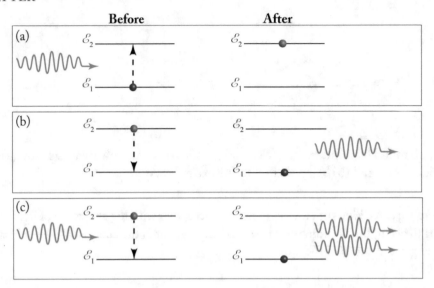

Figure 2.3: Radiative transitions in atoms: (a) photon absorption, (b) spontaneous emission, and (c) stimulated emission.

molecules are considered as the structural units of the material medium. This topic will be discussed in the following section.

What happens if the photons' energy in the light propagating through a medium does not satisfy the relationship (2.1)? In this case, a non-resonant light–matter interaction will occur. The light will interact with the atoms of the medium in a non-resonant fashion, i.e., without inducing any transitions, but it will still experience a delay in the medium acquired in the course of the interaction. We will discuss this in detail in Chapter 3.

Atoms can group to form atomic lattice structures in some materials (see Fig. 2.4a). In some other materials, they can group to form molecules (see Fig. 2.4b), representing another structural unit that we will discuss in the following section.

2.2   MOLECULES AND MOLECULAR TRANSITIONS

Multiple atoms can form a stable molecule by sharing their electrons through electrostatic and quantum mechanical mechanisms, thereby forming bonds holding them together. One of the most common everyday-life examples of the molecular matter is water (H_2O), with molecules comprising two atoms of hydrogen and one atom of oxygen. Like any other substance, water can exist in vapor, liquid, or solid phase. In Fig. 2.4b, we show one of the possible structural arrangements of ice. Each site of its structure consists of a H_2O molecule, containing two hydrogen and one oxygen atoms. The structural unit of a molecular solid is, therefore, a molecule.

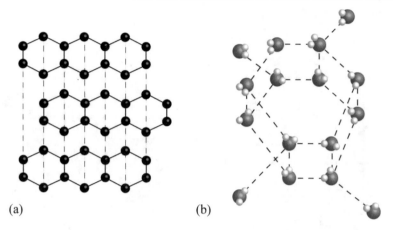

Figure 2.4: Examples of structures of (a) atomic solids (graphite, with carbon atoms shown as circles) and (b) molecular solids (ice molecules arranged in a lattice where each site represents an H_2O molecule with hydrogen atoms shown as small white circles and oxygen atoms shown as red circles). (a) The image is recreated in similarity to https://igcse-chemistry-2017. blogspot.com/2017/07/150-explain-how-structures-of-diamond.html. (b) The image is recreated in similarity to https://www.learner.org/courses/chemistry/text/text.html?dis=U&num= Ym5WdElURS9OQ289&sec=Ym5WdElUQS9OQ289.

The dynamics of molecules are rather complex, reflected in their more sophisticated energy level structures compared to the atomic energy structure we discussed in the previous section. Because molecules vibrate and rotate, they exhibit two additional types of resonances. The first type, *vibrational resonances*, arises from the vibrations of the molecular bonds occurring as a consequence of simultaneous motions of the molecules themselves and the electrons within the atoms that form the molecules. As a result of these vibrations, the energy levels associated with the electrons, forming the molecules, split into multiple sub-levels, corresponding to different vibrational energy states that could be occupied by an electron. In Fig. 2.5a, the energy levels associated with the electrons are schematically represented by thick black solid lines, while the vibrational levels are shown with thinner grey solid lines. Each electron level has a set of the vibrational sub-levels. The resonant frequency of the vibrations (the frequency difference between two vibrational levels within a single electron energy level) falls within the range of 10^{13}–10^{14} Hz. This is about an order of magnitude smaller compared to optical frequencies associated with electronic transitions of atoms (the transitions occurring between different energy levels associated with the electrons). The second, additional type of resonances in molecules, is *rotational resonances* that appear due to molecular rotations. They result in an additional split of the vibrational sub-levels associated with the electronic energy levels into a group of rotational sublevels. They are schematically represented by the grey dashed lines on the diagram shown in

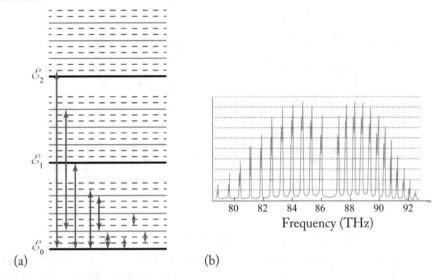

Figure 2.5: (a) Schematic representation of molecular energy levels, showing electronic (thick black solid lines), vibrational (thin grey solid lines), and rotational (grey dashed lines) resonances. The blue, green, and red arrows show possible electronic, vibrational, and rotational transitions, respectively. (b) The infrared absorption spectrum of HCl showing vibrational and rotational molecular resonance peaks. (a) The image is recreated in similarity to http://chem.sci.ubu.ac.th/e-learning/inmr_en/spectroscopy/molecular_energy_level.htm. (b) The figure is the property and courtesy of Rod Nave and was reproduced from the HyperPhysics Project website http://hyperphysics.phy-astr.gsu.edu/hbase/molecule/vibrot.html with permission.

Fig. 2.5a. Typically, a molecular revolution takes a longer time than that required to complete one cycle of the molecular bond vibration. That is why the rotational resonances are characterized by even smaller resonant frequencies falling within the far-infrared range, with the values $\sim 10^{12}$ Hz or 10 THz and lower.

The energy levels associated with molecular resonances have, therefore, the following overall structure (see Fig. 2.5a for a schematic representation). Atoms, comprising the molecules, have their associated electronic energy levels, shown by bold black lines on the diagram. Each electronic level is split into vibrational sub-levels, shown with thinner grey solid lines. Finally, each vibrational sub-level is split into multiple rotational sub-levels characterized by even smaller energy differences (shown with grey dashed lines). The arrows on the diagram show examples of possible transitions between electronic, vibrational and rotational levels. Radiative transitions of all three kinds, discussed in Section 2.1 (see Fig. 2.3 and the associated discussion), can occur in molecules between any two levels of their abundant energy spectra, as long as they are allowed by the laws of quantum mechanics. As a consequence, the transition spectrum of molecules has

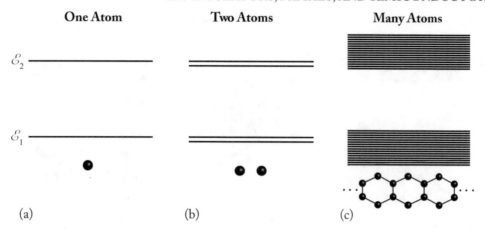

Figure 2.6: A schematic diagram showing (a) a single atom and its selected lower and excited energy states; (b) two atoms interacting with each other and the associated energy splitting of their upper and lower states; and (c) many atoms forming an atomic solid and interacting with each other, and the associated energy splitting of their upper and lower energy states forming two bands of sub-levels.

many peaks associated with all kinds of electronic, vibrational, and rotational resonances (see Fig. 2.5b as an example). Molecular spectroscopy is a separate field of study that deals with matching molecules and fragments of molecules with signatures in their spectra; further discussion associated with this topic is outside the scope of this book.

Further, we discuss energy level structures of dense material media where many atoms (or molecules) are situated close to each other. For the sake of simplicity, we limit our consideration to atomic solids.

2.3   INSULATORS, METALS, AND SEMICONDUCTORS

Let us consider what happens with the ground and excited energy states (some pair of lower and higher atomic levels) when many identical atoms are residing close to each other, which is the case with any atomic solid. Isolated atoms have discrete energy levels (see Fig. 2.6a). If we bring two isolated atoms close to each other, they start interacting with each other. As a result of this interaction, their lower and higher energy levels split due to the electrostatic electron repulsion, and due to the *Pauli exclusion principle*, that requires that no two identical electrons are occupying the same energy level. This is schematically illustrated in Fig. 2.6b.

Let us imagine many such atoms brought together and interacting with each other, as in the case of a solid medium. This interaction of many atoms leads to splitting the ground and excited energy states in many different ways. There are now many sub-levels associated with the

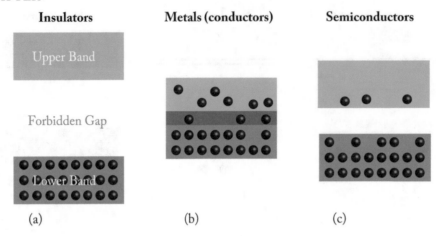

Figure 2.7: Schematic representation of (a) insulators, (b) metals (conductors), and (c) semiconductors with their corresponding energy bands and band gaps.

ground and excited states. These sub-levels are situated very close to each other, which means that the energy difference between two neighboring sub-levels corresponding to the same state (either ground or excited) is tiny. As a result of this energy level splitting, both the ground and excited states evolve into entire *bands* of energy sub-levels; the sub-levels within each band are spaced so closely that each band could be considered as a *quasi-continuum* of energy states (see Fig. 2.6c).

Based on the characteristics of the constituent atoms or molecules, the solid can exhibit markedly different behavior depending on the size of the gap between the lower and higher energy bands. One limiting situation corresponds to a large gap, 10 electronvolts [eV][1] or higher, between the upper and lower energy bands (see Fig. 2.7a). Such a gap is called a *forbidden gap* as there are no energy levels in the gap to be occupied by electrons. In this case, the lower energy band is likely to be occupied by electrons. In contrast, the upper energy band is likely to be empty under the conditions of no external excitation: the thermal excitation where the ambient environment shares its thermal energy with the atomic system is insufficient to excite an electron to the upper energy band. The electrons residing at the lower energy band are held down to the atom by the electrostatic forces existing between these electrons and the nucleus. Such forces decay with the square of the distance from the nucleus and are expected to be significantly weaker for the electrons excited to the upper energy band. Such excited electrons can become mobile (can move freely across the solid). On the other hand, such excitation of the electrons from the lower to higher energy band would require much energy, beyond what traditional excitation

---

[1]The electron volt [eV] is the amount of energy gained by a single electron as it travels across an electric potential of one volt. $1\,eV = 1.6 \times 10^{-19}$ J. It is convenient to use electronvolts in place of joules in the situations where the energies involved are tiny.

methods can offer. As a result, such an atomic solid with the large energy gap between the lower and upper energy bands would act as an insulator as there are no free electrons or mobile carriers capable of producing an electric current. Some examples of insulators used in optics are silica glass, which is used to manufacture optical fibers and components, sapphire, ruby, garnet, used for active laser elements, and many other materials.

Another limiting case corresponds to the situation where the lower and upper energy bands spread sufficiently, such that there is an overlap between their energy sub-levels (see Fig. 2.7b). As a result, it takes minimal effort to excite electrons residing in the lower energy band to the upper energy band. There is an abundance of charge carriers in both energy bands, and these carriers can exhibit mobility. They can produce electric currents as they flow through the surface of the material. Such a material, therefore, exhibits conducting properties, and it is called *conductor*. This is the case with metals. Some metals used in optics are gold, silver, and aluminum, used as coating materials for optical mirrors.

There exists an intermediate situation that bridges the two limiting cases. In some materials, there is a forbidden gap between the upper and lower energy bands. However, the size of this gap is relatively small, not greater than a few electronvolts (see Fig. 2.7c). In this situation, thermal excitations of electrons from the lower to the upper energy band can occur, and there can be a significant population of electrons in the upper level. However, under normal conditions of relatively low ambient temperature (e.g., the room temperature 300 K), this population of free carriers is insufficient to produce a significant current flow. Nevertheless, it is still possible to induce a current flow in such materials using other methods of excitation, e.g., excitation with light. As a result, the material would exhibit the properties of both insulators and conductors. Such materials are called *semiconductors*; they are in the basis of optoelectronic devices that implement current flow to generate light or absorb light to generate current. Some semiconductor materials used in optics include silicon (for integrated optics), germanium (for detectors), III-V semiconductors, which is a group of materials comprised of the elements from the columns III and V of the periodic table. Optics of semiconductors is a separate part of this course; it will be discussed in detail in the concluding book of this series.

2.4 AMORPHOUS AND CRYSTALLINE SOLIDS

As the light propagates through a material medium, it interacts with it. The medium responds to the electromagnetic field of the light, and the nature of the response depends on what kind of medium it is. One example of a physical quantity characterizing the optical response of a material medium is its refractive index n. The optical response of a material medium to incident light can be independent of the polarization of light, i.e., can be the same in all directions regardless of how the light propagated through the medium. However, there are situations where the optical response depends on the propagation direction and polarization of the electromagnetic field associated with the light. The two cases correspond to two different kinds of media that we are going to discuss briefly. For simplicity, we limit our consideration to solid media.

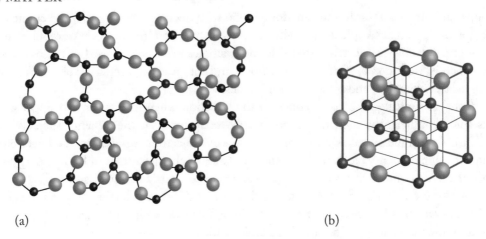

(a)   (b)

Figure 2.8: (a) Internal structure of silica glass. Blue circles represent oxygen while brown circles are silicon atoms. (b) Unit cell of MgO crystal with blue spheres representing oxygen ions and magenta spheres showing magnesium ions. (a) The image is recreated in similarity with https://en.wikipedia.org/wiki/Glass. (b) The image is recreated in similarity with http://www.theochem.unito.it/crystal_tuto/mssc2008_cd/tutorials/surfaces/surfaces_tut.html.

The situation where the propagation direction and polarization of light is of no importance (when the material response is the same in all directions) corresponds to optically isotropic media. Amorphous solids are an example of optically isotropic materials. The molecules or atoms in such solids are distributed randomly without a strictly organized structure. As a result, there is no well-defined periodicity of the atoms in the solid, nor is there any sense of direction in such solids. In this case, direction-independent constants can describe the optical response. Speaking of the refractive index, it is a scalar quantity that has the same value regardless of the light's polarization and propagation direction. Physically, it occurs because, in any direction within the medium, there is an equal number of molecules or atoms per unit length; therefore, the light interacts with an equal number of structural elements of the medium regardless of its orientation. An example of such a medium is silica glass, one of the most abundant materials that are widely used in optics. In Fig. 2.8a, we show a schematic of what the internal structure of silica could look like. The oxygen atoms are shown with blue spheres and the silicon atoms with brown spheres on the figure. By observation, one can conclude that the atoms are distributed randomly. So, there is an approximately equal number of atoms for the light to interact with along any direction within the medium.

Another type of solid, called *crystalline solid* or simply a *crystal*, is comprised of organized arrangements of atoms, forming crystalline structures. In this case, depending on the specific geometry of the unit cell (the unit structural element representing a single cell of the periodic structure), there can be the same or different values of the refractive indices and other optical response parameters in different directions with respect to the crystalline axes. In Fig. 2.8b, we show an example of a unit cell of a magnesium oxide (MgO) crystal. It is an ionic solid comprised of magnesium and oxygen ions held together by strong ionic bonds. The blue spheres represent oxygen ions, while the magenta spheres are magnesium ions. This particular example of a crystalline solid is optically isotropic, which is due to its symmetric cubic structure. The refractive index takes a single value regardless of the polarization orientation with respect to the crystalline axes.

There are, on the other hand, crystalline structures that exhibit *anisotropy* of optical properties: the characteristics of the optical response (optical constants) in such structures are direction-dependent and are, therefore, mathematically described as *tensors* rather than scalar quantities. An example of such a crystal is graphite, shown earlier in Fig. 2.4a. One can conclude by the observation that this crystal has a lower symmetry than MgO: its lattice is elongated along the vertical direction, while it is denser in the horizontal planes of hexagonal formations shown on the diagram. As a result, the refractive index takes different values depending on the polarization of light propagating inside such a crystal. It can no longer be described by a single-value scalar constant. While the mathematical description of light interaction with anisotropic materials is rather complex, there are many useful applications associated with this property of crystals; we will discuss some of those later in Section 3.3 of Chapter 3.

We conclude Chapter 2 by providing a simple description of material media from the optics perspective. For more in-depth learning, we refer the reader to specialized books describing the subject matter at a more advanced level [16, 17]. The next chapter, building on Chapters 1 and 2, will focus on merging light with matter and describing various aspects of their interaction.

CHAPTER 3

Interaction of Light with Matter

In Chapters 1 and 2, we presented an overview of the nature of light and matter. While we introduced and considered them separately, we already started discussing their interaction. Specifically, we introduced the concept of refractive index n of a material medium, the parameter describing its material response and depending on its structural properties. We discussed what kind of influence it has on light propagation through the medium. We helped the reader to develop a sufficient understanding of the properties of light and matter. We are now ready to discuss different aspects of their interaction in more detail, which is the subject of this chapter. The goal is to build a foundation for understanding the principles of operation of optical components and devices.

As light propagates through a material medium, different accompanying phenomena can appear, depending on the propagation conditions. In the simplest case, as we already discussed earlier, it propagates through a homogeneous isotropic medium, experiencing the phase change kz that depends on the refractive index n of the medium. If it is monochromatic light, the refractive index can be treated as a constant, and the impact is merely the slow-down of the light in the medium. If the light has a spectrum of frequencies and comes in the form of optical pulses of finite duration, the refractive index can no longer be treated as a constant in the most general case because of the effect of *dispersion*, or the dependence of the refractive index on the frequency of light. Dispersion plays a crucial role in light propagation through the optical medium and has to be considered in designing optical components and devices. Moreover, it serves as the mechanism of operation in certain optical components and devices. It is an important subject for discussion in this chapter.

Material media respond differently to different colors of light. Some colors, or frequencies, can be far from the medium's resonances. The interaction of such light with the medium is reduced to phase accumulation and slow-down acquired by the light, as well as some possible impact of dispersion. However, if the frequency of light matches one of the material's resonances, it can be absorbed by the optical medium. Absorption represents a source of loss of optical power and can be regarded as a parasitic effect. On the other hand, it can serve as the mechanism of operation for some optoelectronic and optical devices, such as detectors of optical power or optically excited amplifiers and lasers.

Absorption[1] and dispersion are *linear* optical phenomena. The material parameters used to describe these phenomena are field-independent. This means that, when treating dispersion or linear absorption, the material medium responds equally to an optical field of any strength. On the other hand, not all the effects associated with the material response are linear. Some of them are *nonlinear* with intensity-dependent material parameters. In these cases, the outcome of light–matter interaction can strongly depend on the intensity of the applied light. A group of intensity-dependent phenomena is called *nonlinear optical interactions*. Depending on the material medium, there can be a wide variety of such effects. Some of them are accompanied by the change in the color of light, thereby allowing one to access new spectral ranges not attainable by conventional light sources. The field of optics describing such effects is called *nonlinear optics*. To conclude our discussion on the aspects of light–matter interaction, we present a brief overview of nonlinear optics and possible nonlinear optical phenomena, to give the reader a flavor of this colorful, exciting field.

3.1 ABSORPTION

What is happening inside the matter when the light propagates through it? Simplistically, two things are happening: (i) molecules and atoms oscillate with the applied optical field's frequency and (ii) molecules and atoms absorb the light with the frequencies matching their resonances. These two aspects of the interaction of light with matter that we are going to discuss next in more detail cover the whole range of phenomena that can accompany light propagation through a material medium.

3.1.1 OSCILLATIONS OF MOLECULES AND ATOMS UNDER THE APPLIED FIELD

Optical waves exhibit oscillations of electric and magnetic fields in time and space (see Section 1.2.4 of Chapter 1). While molecules and atoms are electrically neutral, they are comprised of elementary charges, protons, and electrons, that interact with the applied electric field. As the wave oscillates, it changes the direction in which its electric field vector is pointing (see Fig. 1.8), and the electrons and protons follow the oscillations of the optical wave.

In Fig. 3.1, we illustrate the oscillations of a water molecule under the applied optical field. The covalent bonding holds a negatively charged oxygen ion and two positively charged hydrogen ions together, bonded into a molecule. As the optical field oscillates, the charges in the molecule follow the sign of the electric field, becoming displaced. Consequently, a temporary spatial charge separation, oscillating with the same frequency as that of the optical field, is induced in the molecule. This charge separation is called *dipole moment*, and it is described as

$$\mathbf{p} = q\mathbf{x}, \tag{3.1}$$

---

[1]Here, we specifically imply *linear absorption* that occurs at the level of single photons. There exists also *nonlinear absorption*, which entails simultaneous absorption of two or more photons (see Section 3.3.3, "Nonparametric Nonlinear Optical Processes").

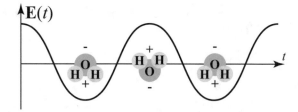

Figure 3.1: Oscillations of a water molecule under the applied optical field.

where q is the charge and $\mathbf{x}$ is the vector characterizing its deflection from the equilibrium position.

The material medium comprises many molecules and atoms, and all of them oscillate with the applied optical field, acquiring dipole moments. As such, the medium can be viewed as a collection of dipoles with dipole moments $\mathbf{p}$. Their oscillation with the frequency ω of the applied optical field results in the periodic change of charge displacement within a dipole, following the law

$$\mathbf{x}(t) = \mathbf{x_0} e^{-i\omega t} + \text{c. c.} \tag{3.2}$$

Here, $\mathbf{x_0}$ is a vector quantity describing the amplitude of the charge's deviation from its equilibrium position. Following Eq. (3.1), the molecule's dipole moment oscillates with the applied optical field.

The dipole moment oscillation under the applied optical field is typically characterized in terms of the average quantity, the dipole moment per unit volume or the *polarization* of the medium $\mathbf{P}$. It is an essential characteristic that carries information about the applied field and the medium's material response on average. The polarization of the medium can be related to the dipole moment of its molecules or atoms as

$$\mathbf{P} = N\mathbf{p}, \tag{3.3}$$

where N is the molecular or atomic density (the number of molecules or atoms per unit volume).

3.1.2   OBJECT COLORS

When a beam of white light interacts with an object that is *absorptive* (has its resonance frequencies coinciding with some spectral components within the beam), the object can partially or entirely eliminate a specific color, or set of colors, from the beam. For example, let us imagine that we have a collimated beam of white light (effectively containing all colors of the rainbow) that falls onto an atomic medium with the resonances in the blue, green, and purple spectral windows. If the resonances are sufficiently strong, and there is a sufficiently large number of atoms interacting with the beam of light, the medium can eliminate, or absorb, the blue, green, and purple components from the propagating beam of light, allowing the rest of the spectral components to traverse it.

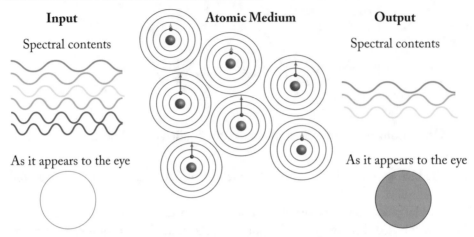

Figure 3.2: Demonstration of the effect of absorption at the atomic level. A collection of atoms has multiple resonances responsible for absorbing different colors from the spectrum of a white light beam. The input spectral content and the way the spectral components blend and appear to a human eye are shown on the left-hand side. The output spectral content and how the transmitted light appears to an eye are shown on the right. The center of the figure schematically shows the atoms' collection, which features different resonant frequencies, absorbing green, blue, and purple.

This example is illustrated in Fig. 3.2. The spectral content of the incident light is schematically represented by different colors on the left-hand side of the figure. The way the frequencies blend and appear to a human eye is shown in the bottom-left corner (a white circle schematically representing the cross-section of the incident beam). The atomic medium, shown in the central part of the figure, is schematically represented as a collection of atoms exhibiting multiple resonances. This representation shows multiple options for an electron in an atom to get excited: by absorbing a green, blue, or purple photon. The spectral content of the radiation escaping from the medium (traversing it) is shown on the right-hand side, with the effective color of the output beam of light shown at the bottom-right. The residual spectral components, red, yellow, and orange, combine to appear as orange.

Another way to demonstrate this concept (less schematic but more realistic) is shown in Fig. 3.3. A beam of white light comprising a mixture of red, green, and blue (the primary colors) interacts with an object shaped like a rectangular bar. The object absorbs blue only. The figure demonstrates the color of light as a human eye perceives it before and after its interaction with the medium. Before the interaction, the light appears to be white as it contains a mixture of red, green, and blue. After the interaction, the white light loses its blue component, while the mixture of red and green remains intact. The mixture of the two remaining primary colors

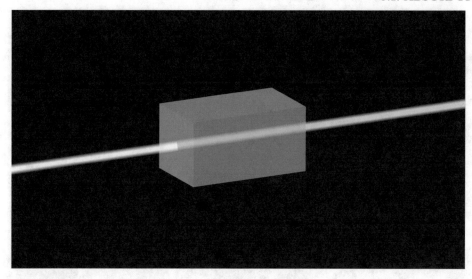

Figure 3.3: Demonstration of absorption of blue spectral component from a beam of white light.

produces yellow. That is why the objects absorbing blue appear yellow as they scatter or transmit the residual light.

The property of objects to absorb specific spectral components (wavelengths) selectively has an application in coloring. In essence, an object that appears in a particular color absorbs its complementary color. This is well demonstrated by a fun experiment with gummy bears of different colors (see Fig. 3.4). The experiment requires six gummy bears, three green and three red, and two laser sources, green and red. On the left-hand side of the figure, green laser light interacts with three red (top) and three green (bottom) gummy bears placed in a row to maximize the effect. When the green light is applied to the red gummy bears, it barely makes it past the first bear: most of it gets absorbed. Nothing reaches the third bear. When the green light interacts with the green gummy bears, it penetrates through them without getting absorbed. The light loses some power on scattering because gummy bears are made of gelatine, a pure natural protein with very large molecules (the molecular length is 300 nm, which is comparable to the wavelength of light).[2] The right-hand side of the figure shows a similar experiment with red light illumination. The green gummy bears (top) absorb red light, while the red ones transmit and scatter it.

An obvious question arises: how does the above observation agree with the color diagram shown in Fig. 1.17? Red color there stands opposite to cyan, not green. One should bear in

[2]When the light interacts with the grains of the size of a wavelength, it experiences *Mie scattering* (see the article   http://www.atmo.arizona.edu/students/courselinks/spring08/atmo336s1/courses/fall13/atmo170a1s3/1S1P_stuff/ scattering_of_light/scattering_of_light.html for more information).

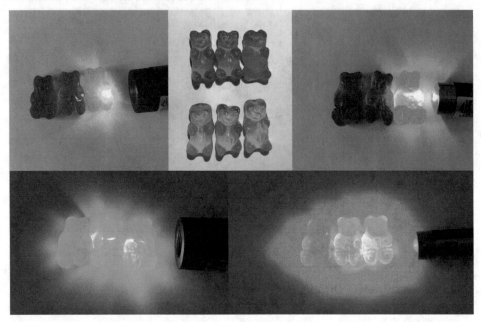

Figure 3.4: A fun experiment with gummy bears aimed at demonstrating interaction of light with colored objects. The top-center portion shows gummy bears under test (red and green). The left-hand images show the application of a green laser light to the three red (top) and three green (bottom) gummy bears. The right-hand images show the application of a red laser light to three green (top) and three red (bottom) gummy bears.

mind that the situation represented in Fig. 1.17 is simplistic: there we deal with "pure red," which has {1, 0, 0} in terms of the fraction of {red, green, blue} (where one corresponds to the full intensity of the color in the mixture), and its opposite color (cyan) has {0, 1, 1}. There are tons of red and green shades that appear to an eye as red and green, respectively. For example, a red tomato can have {1, 0.4, 0.3} as its red, green, and blue component fractions. Depending on the specific shade associated with the red and green, one can find some range where red and green appear as complementary colors. Figure 1.17 supports the above claim by allowing one to observe complementary mixed colors. The color that stands between red and magenta (obtained by mixing those two) still appears as red. Its complementary color, obtained by mixing green and cyan, appears green. Of course, they are quite different from pure red {1, 0, 0} and pure green {0, 1, 0}. A similar situation occurs when orange appears as a complementary color to some shades of blue (see the diagram).

3.1.3   ABSORPTION COEFFICIENT

All optical media are absorptive in a particular spectral range where they have resonances associated with their electronic or molecular transitions (see Chapter 2, Sections 2.1 and 2.2). They can absorb light with the wavelengths matching one (or several) of their resonance frequencies. At the same time, optical materials behave as transparent to the light with the frequency that is far from their resonances. A material medium thus responds differently to different frequencies depending on their proximity to its resonances. This frequency-dependent material response is characterized by a complex material parameter, *optical susceptibility* $\chi(\nu)$, which is a function of frequency[3]:

$$\chi(\nu) = \chi'(\nu) + i\chi''(\nu), \tag{3.4}$$

where

$$\mathrm{Re}[\chi(\nu)] = \chi'(\nu), \tag{3.5a}$$

and

$$\mathrm{Im}[\chi(\nu)] = \chi''(\nu). \tag{3.5b}$$

Later in this section, we will see that the susceptibility having a non-zero imaginary component $\chi''(\nu) \neq 0$ at certain frequencies indicates the presence of material absorption. For the frequencies of light lying far away from the material resonances, $\chi''(\nu) \approx 0$, and the susceptibility becomes real: $\chi(\nu) \approx \chi'(\nu)$.

The susceptibility is a characteristic describing the material response of the medium. It has a specific value, a specific frequency dependence $\chi(\nu)$, and a specific set of resonance frequencies, all unique to the material. It is the susceptibility that dictates the value of the polarization **P** induced in a material medium in response to the applied optical field strength, characterized by its electric field vector **E**:

$$\mathbf{P} = \epsilon_0 \chi \mathbf{E}. \tag{3.6}$$

The reason why the polarization is a vector quantity is that the dipole moment has a specific direction that depends on the polarization of the applied optical field: the direction of the charge separation induced by the field is dictated by the field's polarization, as demonstrated in Fig. 3.1.

The parameter ϵ_0 entering Eq. (3.3) is the *dielectric permittivity of the vacuum*. It is a fundamental constant with the value $\epsilon_0 \approx 8.85 \times 10^{-12}$ [F/m] (farad per meter), which has the meaning of a distributed capacitance of the vacuum, or its capability to permit an electric field. It is related to another fundamental constant that we introduced earlier in the book, the speed of light in vacuum, as

$$c_0 = \frac{1}{\sqrt{\epsilon_0 \mu_0}}. \tag{3.7}$$

---

[3]In an anisotropic medium, $\overleftrightarrow{\chi}$ is a tensor quantity, depending on the orientation of the optical field concerning the medium's internal structure. Here, for simplicity, we consider a material medium that can be described by a scalar function $\chi(\nu)$.

Equation (3.7) contains another fundamental constant, $\mu_0 \approx 12.57 \times 10^{-7}$ [H/m] (henry per meter), called *magnetic permeability of the vacuum*. It represents a measure of resistance associated with forming a magnetic field in vacuum.

In addition to the vacuum values of the dielectric permittivity and magnetic permeability, it is imperative to introduce their material (non-vacuum) counterparts, *dielectric permittivity* and *magnetic permeability*. They have similar physical meanings but describe a material medium. Dielectric permittivity ϵ of a medium is the proportionality coefficient between the electric field associated with light and the *electric displacement* **D** of the charges in the medium, or *electric induction*, induced by the electric field:

$$\mathbf{D} = \epsilon \mathbf{E} = \epsilon_0 \epsilon_r \mathbf{E}. \tag{3.8}$$

It has a unique value for each optical material. The quantity

$$\epsilon_r = \frac{\epsilon}{\epsilon_0}, \tag{3.9}$$

called *relative permittivity*, is the dimensionless dielectric permittivity that characterizes the material properties: ϵ has the same units as ϵ_0, which is expressed in [F/m]. The convenience of using ϵ_r over ϵ will become evident when we relate them to the refractive index of the material.

Magnetic permeability μ of a material medium is the proportionality coefficient between the magnetic field **H** of the applied light wave and the magnetic induction **B**, which is the measure of the induced magnetic field in the medium:

$$\mathbf{B} = \mu \mathbf{H} + \mu \mathbf{M}. \tag{3.10}$$

Here **M** is the *magnetization* of the medium, the vector quantity characterizing the density of permanent or induced magnetic dipole moments, which is an equivalent of the electric polarization, but in the context of magnetic properties of the medium. Unlike polarization, magnetization can only exist in magnetic media. While each medium generally has unique values of μ and **M**, in the majority of cases of interest in optics, $\mu = \mu_0$, and **M** = 0 for optical materials. We restrict our consideration to these cases as other cases are rather exotic and are mostly related to artificial nanostructures involving metals, which is beyond the scope of this introductory course. Based on this argument, we adapt

$$\mathbf{B} = \mu_0 \mathbf{H} \tag{3.11}$$

for the rest of the book series.

We now have several parameters describing the material response: the susceptibility, absolute and relative dielectric permittivities. They are related to each other as

$$\epsilon(\nu) = \epsilon_0 [1 + \chi(\nu)] \tag{3.12}$$

and

$$\epsilon_r(\nu) = 1 + \chi(\nu). \tag{3.13}$$

Since the susceptibility is, in general, complex and frequency-dependent, the dielectric permittivity is also complex and frequency-dependent. The same happens with the wavenumber $k(v)$ in the medium that now acquires an imaginary part as well:

$$k(v) = \frac{\omega}{c_0}\sqrt{\epsilon_r(v)} = k_0\sqrt{\epsilon_r(v)}. \tag{3.14}$$

Under the conditions of non-resonant light–matter interaction, $\sqrt{\epsilon_r(v)} = n$, Eq. (3.14) becomes real and reduces to Eq. (1.5) that we introduced in Chapter 1.

Using Eqs. (3.12) and (3.4), we can rewrite Eq. (3.14) as

$$k(v) = k_0\sqrt{1 + \chi(v)} = k_0\sqrt{1 + \chi'(v) + \chi''(v)}. \tag{3.15}$$

Equation (3.15) reveals the complexity of the wave number, coming from the complexity of the material response parameters. We can rewrite the expression for $k(v)$ as

$$k(v) = \beta + i\frac{\alpha}{2} \tag{3.16}$$

to represent it as a sum of its real and imaginary parts.

Let us now substitute Eq. (3.16) into Eq. (1.17) for a plane wave propagating along Z-axis to demonstrate the physical meaning of the quantities α and β. We obtain

$$U(z) = U_0 \exp(-i\beta z)\exp\left(-\frac{\alpha}{2}z\right). \tag{3.17}$$

The term $\exp(-i\beta z)$ is similar to the term $\exp(-i\mathbf{k}\cdot\mathbf{r})$ of Eq. (1.17), which describes the phase shift acquired by the wave as it propagates. The real quantity β plays the role of the wave number k in the sense that it describes the phase evolution of the wave as it propagates through an optical medium. The parameter β is called the *propagation constant*. The second term in Eq. (3.17), $\exp[-(\alpha/2)z]$, is an exponent with a negative and real factor. If $\alpha > 0$, this factor describes the attenuation of the wave as it propagates through the optical medium, because the term $\exp[-(\alpha/2)z]$ decreases with the increase of the distance z the wave travels inside the medium. This term thus describes the loss encountered by the field as it propagates through the medium, or the loss of the electric field strength due to the absorption in the medium. The coefficient α is called *absorption coefficient*.

The expressions (3.16) and (3.17) contain the factor $\alpha/2$. According to Eqs. (1.19) and (3.17), optical intensity can be described as

$$I = |U|^2 = |U_0|^2 \exp(-\alpha z). \tag{3.18}$$

The coefficient α is thus the absorption coefficient with respect to optical power or intensity, while $\alpha/2$ is its electric field counterpart. Both α and β are generally frequency dependent, and are measured in the units of the inverse length: [1/m].

As we discussed above, the complexity of the parameters of material response χ and ϵ stems from the fact that optical media are, in general, absorptive. On the other hand, the media exhibit absorption only in certain frequency ranges where they have resonances. The material response parameters χ and ϵ (ϵ_r) are, thus, complex only in the vicinity of resonances where their imaginary parts are nonzero: $\chi'' \neq 0$ and $\epsilon'' \neq 0$. Away from the resonances, $\chi'' = \epsilon'' = 0$ and χ and ϵ (ϵ_r) become real. When the latter holds,

$$\epsilon_r = n^2, \tag{3.19}$$

and $\sqrt{\epsilon_r} = n$ in Eq. (3.14).

We can summarize what we have learned so far in this section in the following manner. The material response parameters can be subdivided into two groups. The first group of the response parameters includes complex parameters, χ and ϵ (ϵ_r) that are associated with the charge separation or dipole moment (polarization), induced in the medium by the applied optical field. They differ slightly from each other because they serve as proportionality constants in relating the strength of the applied field $\mathbf{E}$ to two different characteristics. The dielectric permittivity enters the relationship between the field and electric displacement of $\mathbf{D}$, which is the measure of the induced charge displacement. The susceptibility relates the electric field to the polarization $\mathbf{P}$ associated with the dipole moment induced in the molecules or atoms of the medium by the applied field. The vectors $\mathbf{D}$ and $\mathbf{P}$ are related as

$$\mathbf{D} = \epsilon_0 \mathbf{E} + \mathbf{P}. \tag{3.20}$$

Substituting Eq. (3.3) into (3.20), and taking into account Eq. (3.8), one can obtain the relationships (3.12) between ϵ and χ that we introduced earlier.

The second group of parameters characterizing the material response contains the refractive index n, the absorption coefficient α, and the propagation constant β. These parameters are real. Similar to χ and ϵ, they also describe the properties of a specific optical medium. On the other hand, unlike χ and ϵ that describe the overall material response of the optical medium to the incident light, they are tied down to describing a specific physical phenomenon accompanying light propagation through the medium, such as absorption and phase accumulation. That is why these parameters do not have to be complex.

3.2 DISPERSION

Let us now focus on another aspect of frequency-dependent optical response, namely: *dispersion*. All optical media exhibit dispersion, which means that the optical response characteristics $\epsilon(\nu)$ and $\chi(\nu)$ are frequency- or wavelength-dependent. This can be understood by tracing an analogy between an optical system where molecules or atoms play the role of electric dipoles in the optical field and a mechanical system where a harmonic oscillator moves under the applied inducing force. The mathematical description is also similar; the basic principles can be adapted from mechanics and translated to optics. Below we present a simple model describing the oscillation of bound charges[4] associated with atoms and molecules in the optical field, and providing the explicit form of the susceptibility $\chi(\nu)$ as a function of the frequency of light.

3.2.1   HARMONIC OSCILLATOR MODEL

The specific function describing the frequency dependence of the susceptibility can be obtained with an excellent precision if one considers the molecules or atoms forming a dielectric material as dipoles oscillating in the applied optical field. We briefly discussed this kind of oscillation at the beginning of the chapter (see the example with a water molecule in Fig. 3.1). Under the applied optical field, atoms inside the optical material behave as driven harmonic oscillators (with the field acting as the driving force). Therefore, the formalism developed in mechanics for describing a driven mechanical oscillator, a pendulum or a spring, can be mapped onto an optical oscillator, which is an atom under the applied optical field. Let us see how.

The equation describing the dynamics of a mechanical driven harmonic oscillator has the form

$$\frac{d^2x}{dt^2} + \gamma \frac{dx}{dt} + \omega_0^2 x = \frac{F}{m}. \tag{3.21}$$

Here, x is the coordinate of the oscillator that varies in response to the time-varying inducing force $F(t)$, γ is the damping coefficient characterizing the resistance and friction that contribute to the attenuation of the oscillator's motions (the damping mechanisms), ω_0 is the resonance angular frequency of the oscillator, and m is its mass.

Now, let us consider a single-resonance atomic system, focusing on a particular electron resonance of atoms in the medium. Here and below, we consider the equations in the scalar form to account for a specific polarization direction. In optics, atoms oscillate with the applied optical field, which now plays the role of the inducing force. The relationship between the electric field associated with a light wave and the force imposed by the field on the electrons in the medium has the form

$$E(t) = -\frac{F(t)}{e}, \tag{3.22}$$

where $-e$ is the charge of an electron. The electrons in the atoms displace with the applied optical field that has the temporal variation

$$E(t) = E_0 \cos(\omega t), \tag{3.23}$$

in accordance with Eq. (1.12). Here, E_0 is the time-invariable amplitude of the wave. The field oscillates at frequency ω. Following the field's oscillations, the charge displacement oscillates at frequency ω as well, according to Eq. (3.2). The oscillation of the displacement x of the charges

---

[4]By *bound charges* we mean the positive and negative charges associated with atoms or molecules that experience separation due to the applied optical field. As these charges are tied down to each other due to the electrostatic attraction, one calls them bound. This definition emphasizes the distinction from free carriers that are mobile and can freely move through a material, like free electrons in metals.

from their equilibrium positions results in the oscillating dipole moment in accordance with $p = -ex$, as given by Eq. (3.1). The oscillating dipole moment acquired by individual atoms of the medium results in the polarization induced in the medium as a whole, given by $P = Np$ (see Eq. (3.3)). Substituting Eqs. (3.3) and (3.22) into the equation for the mechanical harmonic oscillator (3.21) and multiplying both parts of it by $-eN$, we obtain a similar equation describing the oscillations of the polarization in the medium

$$\frac{d^2 P}{dt^2} + \gamma \frac{dP}{dt} + \omega_0^2 P = \frac{Ne^2}{m} E, \tag{3.24}$$

where the mass m is the mass of the bound charge, γ is the damping of the atomic oscillations, and ω_0 is the resonance frequency of the atomic system. Introducing the constant

$$\chi_0 = \frac{Ne^2}{\epsilon_0 m \omega_0^2}, \tag{3.25}$$

we can rewrite Eq. (3.24) in the form

$$\frac{d^2 P}{dt^2} + \gamma \frac{dP}{dt} + \omega_0^2 P = \omega_0^2 \epsilon_0 \chi_0 E. \tag{3.26}$$

The polarization follows the oscillations of the electric field of the incident optical wave, as given by Eq. (3.23), and can be represented as

$$P(t) = P_0 \cos(\omega t), \tag{3.27}$$

where P_0 is time-invariable part of the polarization. Substituting Eq. (3.27) into Eq. (3.26), we obtain

$$P = \frac{\epsilon_0 \chi_0 \omega_0^2}{\omega_0^2 - \omega^2 + i\gamma\omega} E. \tag{3.28}$$

The polarization P, induced by the optical field E in the medium, depends on the material response, characterized by the susceptibility $\chi(\nu)$, as described by Eq. (3.6). By equating Eq. (3.28) with the scalar form of Eq. (3.6), $P = \epsilon_0 \chi(\nu) E$, and using the relationship $\omega = 2\pi\nu$, we obtain $\chi(\nu)$ in the form

$$\chi(\nu) = \frac{\chi_0 \omega_0^2}{\omega_0^2 - \omega^2 + i\gamma\omega} = \frac{\chi_0 \nu_0^2}{\nu_0^2 - \nu^2 + i\nu\Delta\nu}, \tag{3.29}$$

where $\Delta\nu = \gamma/2\pi$ is the width of the resonance. The real and imaginary parts of $\chi(\nu)$, related to the refractive index and absorption coefficient, can be expressed as

$$\chi'(\nu) = \chi_0 \frac{\nu_0^2(\nu_0^2 - \nu^2)}{[(\nu_0^2 - \nu^2) + \nu^2 \Delta\nu^2]} \tag{3.30a}$$

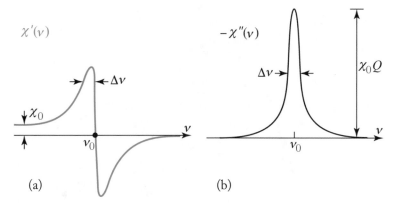

Figure 3.5: Real (a) and imaginary (b) parts of the complex susceptibility $\chi(\nu)$, plotted as functions of frequency. The figure is recreated in similarity to Fig. 5.5-6 from *Fundamentals of Photonics* by Saleh and Teich [8].

and

$$\chi''(\nu) = -\chi_0 \frac{\nu_0^2 \nu \Delta\nu}{[(\nu_0^2 - \nu^2) + \nu^2 \Delta\nu^2]}, \tag{3.30b}$$

respectively. These characteristics are plotted in Fig. 3.5. The parameter $Q = \nu_0/\Delta\nu$ on the graph represents the strength of the resonance ($\chi_0 Q$), i.e., how powerful the absorption is.

The Harmonic Oscillator model is described in the book *Fundamentals of Photonics* by Saleh and Teich [8] (see Chapter 5, Section 5.5.C). Here the author has provided her alternative description to assist less advanced students with understanding this subject matter.

3.2.2 WAVELENGTH-DEPENDENT REFRACTIVE INDEX

The complex susceptibility describes the material response in general. At the same time, its real and imaginary parts are proportional to the refractive index and absorption coefficient, representing real physical parameters describing specific aspects of light–matter interaction. Let us explore how the dispersion of the optical susceptibility translates into the wavelength dependence of the optical medium's refractive index.

One can consider particular optical material to examine its dispersion characteristics. Specifically, one can explore how much its refractive index changes in response to an incremental change in the frequency of light through the entire frequency window from UV to near- or mid-IR. Following this, one can identify regions of strong and weak dispersion. Notably, one can remark that the refractive index exhibits a steeper change as a function of frequency in the resonances' vicinity. In contrast, it exhibits a relatively flat behavior in the regions away from the resonances. It is commonly the case that the regions of strong dispersion are accompanied by

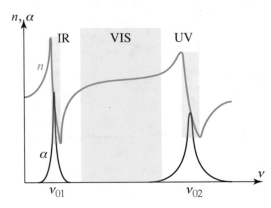

Figure 3.6: Schematic representation of the refractive index (red lines) and absorption (black lines) of a typical dielectric optical material, exhibiting resonances in the UV and IR spectral ranges. The regions of anomalous dispersion are shown with grey shaded areas.

relatively strong absorption, while the regions with flat or weak dispersion overlap with low-loss optical windows.

In Fig. 3.6, we show an illustration of a typical dispersion curve of a dielectric optical material where the electronic resonances occur in the UV spectral range, while the resonances associated with molecular and lattice vibrations fall within the IR spectral window. The corresponding resonance frequencies are marked as v_{01} and v_{02}, respectively. It can be seen from the figure that the refractive index (solid red line) exhibits an increase with the frequency everywhere except right at the resonances where it experiences a dramatic decrease. The regions where it grows with the frequency are thus called the regions of *normal dispersion*, while the regions where it exhibits decrease are the regions of *anomalous dispersion* (shown with grey shaded areas on the figure). The absorption coefficient (solid black line) is only nonzero in the vicinity of the resonances, and it is nearly zero elsewhere. The dispersion curve of the refractive index is relatively flat between the resonances (the region of low dispersion). At the same time, it becomes steeper close to the resonances (the regions of high dispersion and non-negligible absorption).

It can be seen from the schematic represented in Fig. 3.6 that the absorption coefficient is nearly zero in the spectral window away from the resonances. This window is called *the window of transparency*. Figure 3.7 demonstrates the windows of transparency of some materials used in optics for manufacturing optical components and integrated optical devices. To give the reader an idea of how these materials appear to a human eye, we highlight each material's transparency window by its perceived color. Most of the materials are dielectric and transparent in the visible spectral range, except for semiconductors such as zinc selenide, silicon, and germanium. From the schematic, zinc selenide absorbs shorter-frequency visible light, which results in its appearance in the complementary color, which is orange (green and

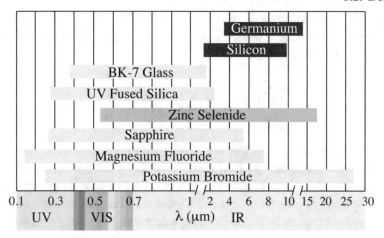

Figure 3.7: The windows of transparency of some dielectric and semiconductor materials used in optics. The color in which the transparency windows are highlighted corresponds to the actual color of the materials, as seen by the eye.

blue spectral components are excluded, leaving behind a blend of yellow, orange, and red). With silicon and germanium, the situation is different because these materials are not transparent in the visible. As a result, they appear brownish-black to a human eye. All the transparent materials on the diagram (shown with greyish color) are dielectrics; the most commonly used ones in optics are BK-7 glass and fused silica. Due to their abundance, low-cost and well-established manufacturing processes, they are used in producing optical components such as lenses and prisms. In the case where an extended transparency window is desired, one can consider magnesium fluoride or potassium bromide for producing optical components: they are transparent from shorter-wavelength UV to mid-IR. One can learn more about the transparency windows and properties of a broader range of optical materials from a variety of databases available from the companies that manufacture optics. Here are some examples: Edmund Optics https://www.edmundoptics.com/knowledge-center/application-notes/optics/understanding-optical-windows/, Thorlabs https://www.thorlabs.com/newgrouppage9.cfm?objectgroup_id=6973, general resource available through Photonics Marketplace https://www.photonics.com/Articles/Transmission_Ranges_for_Optical_Materials/a25105.

Finally, the whole point of deriving Eq. (3.29) and providing the explicit form of optical susceptibility as a function of frequency was to relate this form to a more practical expression for the refractive index as a function of wavelength. In manufacturing optical components, it is imperative to have accurate information about the dispersion properties of the materials from which these optical components are made. Specifically, one wants to know the explicit expression for the refractive index as a function of wavelength in a broad spectral window covering the operation range of the optical component to be manufactured. To serve this goal, there is a practical approach to retrieving the refractive index as a function of wavelength from experimental measurements and fitting it to an empirical model called *Sellmeier equation*. The general form of Sellmeier equation is given as

$$n^2(\lambda) = a + \sum_i b_i \frac{\lambda^2}{\lambda^2 - c_i}, \tag{3.31}$$

where the summation occurs over all the resonances relevant to a specific spectral window of interest (that contribute significantly to the characteristics of the optical response in this window). Here the refractive index is considered as a function of wavelength (the most practical representation in optics), and the parameters a, b_i, and c_i, where the subscript i refers to a specific contributing resonance, can be found from experimentally measured values of the refractive index, obtained for a limited number of wavelengths. Once the refractive index measurements are performed using one of the standard measurement methods, one typically fits the Sellmeier equation to the measured values, which allows one to obtain the parameters for Eq. (3.31). This procedure allows one to predict the value of the refractive index at any wavelength. Normally, considering only two or three terms in the summation in Eq. (3.31) when performing the fit yields sufficient precision. Since the refractive index is the fundamental property of optical materials, the Sellmeier coefficients a, b_i, and c_i are readily available for many optical materials. They can be found in one of the materials databases accessible online (see, for example, https://www.rp-photonics.com/sellmeier_formula.html, https://refractiveindex.info/). There are also additional resources on fitting the experimental refractive index data with the Sellmeier equation (see, for example, [18]).

Before covering the influence of dispersion on the optical pulse propagation, we remark that one of the manifestations of the refractive index's dispersion is the ability of some transparent optical components to deflect light of different colors by different angles. We have provided such an example in Chapter 1, where we show an optical prism (see Fig. 1.16). The fact that the prism exhibits different refractive indices for different colors of light results in a beam of white light splitting into multiple colors representing its spectral content. In such a way, one can retrieve a rainbow of colors at the output of the prism, as illustrated in the figure.

3.2.3   OPTICAL PULSE PROPAGATION THROUGH A DISPERSIVE MEDIUM

The fact that the refractive index for any material medium is wavelength-dependent results in the wavelength-dependent speed of light in the medium (by virtue of Eq. (1.2)). Wavelength-dependent speed of light in the medium implies that the time necessary for a light wave to traverse the medium will depend on its color. In the spectral regions with normal dispersion, the speed of light of the longer-wavelength spectral components is faster than that of the shorter-wavelength components. The opposite is valid for the regions of anomalous dispersion.

The consequences of the wavelength-dependent speed of light are significant for broadband radiation, i.e., for the light with a relatively broad frequency spectrum containing a range of colors. An extreme example of this is the spectrum of the sun (see Fig. 1.15). A more practical laboratory example where the dispersion of the speed of light plays a significant role represents a short optical pulse with a finite-width spectrum.

In order to quantify the impact of dispersion on the optical pulse propagation, there exist various characteristics, some of which we are going to consider next. First, the frequency-dependent refractive index does not only lead to the frequency-dependent speed of light. It also leads to the fact that the propagation constant β is frequency-dependent as well, by virtue of Eq. (1.5) that takes the form $\beta = nk_0$ in line with the discussion presented in Section 3.1. The first parameter characterizing the influence of dispersion on light propagation is *group delay per unit length* $[\text{m}^{-1}\text{s}]$, defined as

$$\beta_1 = \frac{d\beta}{d\omega}. \tag{3.32}$$

The physical meaning of this characteristic is as follows. Different colors can be viewed as different spectral groups (groups of frequencies). Because of dispersion in an optical medium, different spectral groups propagate through it with different velocities and experience different delays, which explains the terminology *group delay*. The group velocity, characterizing the velocity of propagation of a particular group of frequencies, is defined as

$$v_g = \frac{1}{\beta_1}. \tag{3.33}$$

To account for the difference in the propagation velocity experienced by different spectral groups, there is another parameter, called *group velocity dispersion* (GVD). It is defined as

$$\beta_2 = \frac{d^2\beta}{d\omega^2}. \tag{3.34}$$

GVD is an important parameter, often used in practice, for example, to characterize pulse propagation in optical fibers. The physical units in which GVD is quantified are $[\text{s}^2/\text{m}]$. Technically, the parameters β_1 and β_2 are merely the expansion coefficients in

$$\beta(\omega)L = \beta(\omega_0)L + \beta_1(\omega_0)(\omega - \omega_0)L + \beta_2(\omega_0)(\omega - \omega_0)^2 L + \cdots, \tag{3.35}$$

about the frequency ω_0, where L is the length of the dispersive medium. They represent different orders of dispersion. The expansion can continue to higher orders of dispersion, which is sometimes necessary in practice. Nevertheless, the most commonly used parameters are the two lowest-order dispersion coefficients β_1 and β_2.

Another related parameter, also frequently used in quantifying the dispersion characteristics of an optical medium, is called *dispersion parameter*. It is a wavelength derivative of group delay:

$$D = \frac{d}{d\lambda}\left(\frac{1}{v_g}\right) = -\frac{2\pi c_0}{\lambda_0^2}\beta_2. \tag{3.36}$$

The units in which the parameter D is quantified are $[s/m^2]$. However, when one refers to a specific case, such as optical fibers, it is typical to express D in the units of $[ps/(nm\ km)]$. Here ps (picoseconds) refer to a standard duration of optical pulses propagating through the fiber, nm (nanometers) refers to an incremental change in wavelength for which the dispersive properties of the fiber are evaluated, and km (kilometers) to the typical span of the optical fiber. We will revisit the topic of dispersion when discussing light propagation through optical fibers in the next book.

As discussed above, dispersion in an optical medium results in different spectral groups traveling with different velocities through the medium. What implication does it have for an optical pulse? As per the time-bandwidth product relationship (1.36), a relatively short optical pulse should have a relatively broad spectrum. For certainty, let us consider a transform-limited Gaussian pulse with 2 ps temporal FWHM, originating from a pulsed laser. TBP for transform-limited Gaussian pulses is given by Eq. (1.37); it is simply 0.44. From here, we can find the spectral FWHM to be 220 GHz, or, in wavelength equivalent, 1.76 nm at the central wavelength $\lambda_0 = 1.55\ \mu m$.

Half-a-nanometer difference between spectral components forming the spectral content of an optical pulse would not be noticeable by a human eye. Moreover, a significant pulse spread due to a dispersion of the speed of light in the medium would require a long propagation distance in a highly dispersive medium. Nevertheless, the impact of dispersion cannot be ignored in many practical situations where the propagation of such pulses is examined. One such example is the propagation of optical pulses through a fiber, which can be as long as several kilometers. At such a long distance, even a minor difference in wavelengths will result in a significant accumulation of the difference between the time arrivals of two different spectral components. The difference in their time arrivals at the optical fiber output translates into the optical pulse broadening in the fiber. This broadening (additional contribution to the original pulse duration) can be quantified using

$$\Delta T = \frac{d}{d\omega}\left(\frac{L}{v_g}\right)\Delta\omega = L\beta_2\Delta\omega. \tag{3.37}$$

Here, ΔT is the difference in the arrival time of the two most distant spectral components of the optical pulse, having the frequency difference $\Delta\omega$, as the pulse propagates through an optical fiber by the distance L, and β_2, in this case, is the GVD of the fiber. Alternatively, one can write Eq. (3.37) in its wavelength equivalent form as

$$\Delta T = \frac{d}{d\lambda}\left(\frac{L}{v_g}\right)\Delta\lambda = DL\Delta\lambda, \tag{3.38}$$

which is often a more convenient representation in optics. Considering a 2-ps optical pulse with $\Delta\lambda = 1.76$ nm and the central wavelength $\lambda_0 = 1.55\ \mu m$, propagating through $L = 10$ km of optical fiber, characterized by the value of the dispersion parameter $D = 20$ ps/(nm km), we are looking at the pulse spread $\Delta T = 20$ ps/(nm km)$\times 10$ km$\times 1.76$ nm≈ 350 ps. In comparison to the original pulse duration of 2 ps, it is a huge value! The result will vary depending on the central wavelength of the optical pulse.

This particular example shows that even a narrow-spectrum pulse can suffer a significant temporal broadening when it propagates through a dispersive medium for a long time. For even shorter optical pulses, with the temporal FWHM 100 fs, the wavelength spectrum of the pulse will constitute around 10 nm, large enough to be noticed by a human eye. If we compare, for example, 570 nm and 580 nm, we notice a significant change in the color of light from green-yellow to yellow. For more information, one can refer to the chart shown in Fig. 1.6. From the point of dispersion, it means that to observe a significant pulse broadening, it takes less than propagating the pulse through a 10-km-piece of optical fiber. It is sufficient to let the pulse to propagate through an optical medium with a moderate amount of dispersion for a relatively short distance. That is why it is important to consider all the factors, such as the value of the dispersion parameters of the material medium at the specific wavelength range, the length of the propagation of the optical pulse through the medium, and the pulse duration and spectral width when deciding whether the effect of dispersion can or cannot be neglected.

The manifestation of the pulse propagation through a dispersive medium is not limited to the temporal spread of the optical pulse, as discussed in the above example. A typical phenomenon accompanying such spread stems from the fact that different groups of frequencies within the pulse's spectrum travel through the medium at different velocities. The temporal broadening of the pulse indicates the difference in their arrival times at the medium's output. In this case, it is typical to observe a frequency redistribution across the pulse's temporal envelope or *optical chirp*. This is schematically demonstrated in Fig. 3.8. If the original pulse was transform-limited before entering the dispersive medium (see Fig. 3.8a), after propagating through the medium with a significant dispersion, it comes out broadened and chirped. If the medium is characterized by normal dispersion, the pulse will appear as having a red nose and purple tail (see Fig. 3.8b). It means that its longer-wavelength spectral components will be ahead, appearing earlier in time, of its shorter-wavelength (and higher-frequency) components that appear later in time, as shown in the figure. If, on the contrary, the medium exhibits anomalous dispersion, the pulse will come out with a purple nose and red tail (see Fig. 3.8c). Of course, this example is figurative, and the colors are used merely to illustrate the effect of dispersion most clearly. The same is valid for IR pulses where the various frequency components within the spectrum of the pulse can differ significantly from each other (the spectrum can be very broad). However, none of them fall within the visible spectral range, and we will not see the actual red (or purple) nose and purple (or red) tail in the optical pulse as it comes out from a dispersive medium. There will, however, be a significant spectral chirp within the pulse.

In order to mathematically describe the modifications introduced to an optical pulse by a chirp, let us recall that transform-limited pulse has its power spectral envelope represented as

$$P(t) = P_0 \exp\left(-\frac{t^2}{\tau_0^2}\right). \tag{3.39}$$

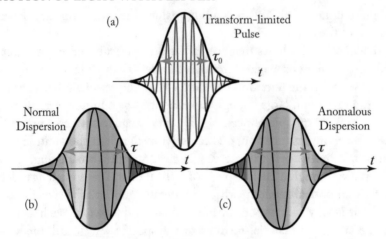

Figure 3.8: (a) Transform-limited optical pulse where each instant of time within its temporal envelope contains all the spectral components. (b) Positively chirped optical pulse suffered propagation through a medium with normal dispersion. (c) Negatively chirped optical pulse that acquired its chirp as a result of propagation through a medium with anomalous dispersion. In both (b) and (c), $\tau > \tau_0$ is the new duration of the pulse's temporal envelope after it suffered the dispersion broadening.

Here, the function $P(t)$ describes time variation of the optical power within the pulse, P_0 represents the peak power, and τ_0 is the temporal half-width at $1/e$ from the peak power value of the pulse (see Chapter 1, Section 1.7). We introduce the subscript "0" to the temporal pulse width parameter τ to emphasize that it is the minimal, transform-limited duration of the pulse before the chirp. If we take the same optical pulse but after it experiences some chirping, the mathematical expression used to describe its temporal power envelope will evolve into

$$P(t) = P_0 \exp\left[-(1 + i\kappa)\frac{t^2}{\tau_0^2}\right] \tag{3.40}$$

to account for the effect of chirping. The chirp parameter κ (a Greek letter "kappa" is used to designate this quantity) can be either positive, which corresponds to the impact of normal dispersion, or negative, representing the effect of anomalous dispersion.

When studying the effect of dispersion for the first time, it is easy to get confused with the following two facts. We list them below in order to facilitate understanding of this subject.

- Dispersion affects the optical pulse duration and can introduce frequency chirp, but it does not modify the pulse's spectrum. It cannot generate new spectral components, nor can it eliminate some frequencies from the spectrum of the pulse, because it is a linear optical phenomenon.

- Transform-limited optical pulse has the shortest duration. There is no chirping that can shrink the pulse duration to make it shorter. Any chirp (either positive or negative) will necessarily broaden the transform-limited pulse. The chirp sign indicates which spectral components will be ahead at the front of the pulse, and which ones will be lagging behind.

This concludes light–matter interaction in the linear optical regime. We refer the reader to Chapter 5, Section 5.5 of *Fundamentals of Photonics* by Saleh and Teich [8] for more advanced coverage of the topics of absorption and dispersion.

3.3   NONLINEAR OPTICS

Nonlinear optics reveals a plethora of colorful phenomena where the frequency of light can change by virtue of light–matter interaction. One example of such frequency conversion is the *second-harmonic generation* (SHG), where the frequency of light ω_1, falling onto an optical medium, doubles as it interacts with the medium in a nonlinear manner, producing a new frequency component $\omega_2 = 2\omega_1$. The incident light in this context is called the *fundamental* radiation. This section aims at explaining the principles of nonlinear optics in simple terms. There is a seminal *Nonlinear Optics* book written by Robert W. Boyd [19] that the reader can refer to for more in-depth studying.

In Fig. 3.9, we show the experiment that illustrates this phenomenon [20]. Part (a) of the image displays a photograph of the optical setup used to conduct the SHG experiment. A laser, emitting red light with the wavelength $\lambda_1 = 800$ nm, was used as the source of the incident fundamental radiation. A nonlinear crystal (inserted into a circular rotating mount, shown in the picture) was used to convert the fundamental radiation into $\lambda_2 = 400$ nm blue second-harmonic light: frequency doubling corresponds to shortening the wavelength twice. The prism was used to spatially separate the residual fundamental and newly generated second-harmonic beams, coming out of the nonlinear crystal. As a result, it was possible to project the two beams onto a paper screen so that they did not spatially overlap. A close-up view of the beam projection onto the screen is shown in Part (b) of the picture. Finally, Part (c) contains a photograph of the beam paths, taken in the dark. It shows the same setup components, labeled, the fundamental red beam propagating toward the nonlinear crystal, and the change in color as the two beams, the fundamental and SHG, come out of the crystal and propagate through the prism toward the screen. One can see a mixture of blue and red originating from the nonlinear crystal. In such a way, starting with red light, we converted part of it into blue light through nonlinear optical interactions in the crystal. Let us look into the nature of nonlinear optical response in materials to observe how light–matter interaction can modify the original light beam and produce new colors.

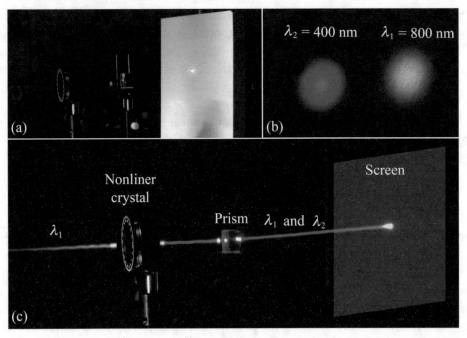

Figure 3.9: Second-harmonic generation experiment. (a) A photograph of the experimental setup, featuring a nonlinear crystal and a prism, with the fundamental laser and generated second-harmonic beams projected on the screen. (b) A close-up view of the fundamental ($\lambda_1 = 800$ nm) and second-harmonic ($\lambda_2 = 400$ nm) beams projected on the screen. (c) A photograph of the light paths, taken in the dark. The key setup components are marked: the nonlinear crystal, prism, screen. The figure was derived from experimental shots the author made for her article on "Frequency Doubling" published in the *Encyclopedia of Nonlinear Science* [20].

3.3.1   NONLINEAR POLARIZATION

In Sections 3.1 and 3.2, we considered the aspects of light–matter interaction in the linear optical regime where the polarization induced in an optical medium scaled linearly with the strength of the incident optical field. All the accompanying phenomena, such as absorption and dispersion, are thus called *linear optical effects*. The Harmonic Oscillator model works well to describe these effects, as long as the optical field's strength is sufficiently low. In this section, we go beyond the linear optical regime and describe the instances where the optical field's strength is high enough to break the limits of the Harmonic Oscillator model.

   If the reader ever had a chance to play with a mechanical oscillator (such as a pendulum or a spring), they might have noticed that the oscillations are periodic as long as the inducing force is moderate. What happens if one applies a very strong external force? The harmonicity of the oscillator's motions will break; it will exhibit more complex motion patterns than those described by a regular sinusoidal function. This behavior stems from the fact that additional harmonics appear in the oscillation spectrum of the pendulum or spring.

When it comes to optics, the same situation arises when the inducing force, which is proportional to the applied optical field strength, is very high. In this situation, the polarization induced in the medium in response to the field can no longer be described by Eq. (3.6) because it acquires components that scale nonlinearly with the strength of the field. A commonly accepted approach in this case is to perform a power series expansion of the polarization with respect to the applied optical field strength:

$$\mathbf{P} = \epsilon_0 \overset{\leftrightarrow}{\chi} \mathbf{E} = \epsilon_0 \overset{\leftrightarrow}{\chi}^{(1)} \cdot \mathbf{E} + \epsilon_0 \overset{\leftrightarrow}{\chi}^{(2)} : \mathbf{EE} + \epsilon_0 \overset{\leftrightarrow}{\chi}^{(3)} \vdots \mathbf{EEE} + \cdots . \tag{3.41}$$

Optical susceptibility $\overset{\leftrightarrow}{\chi}$ in Eq. (3.41) has a tensor form for the reason that nonlinear optical interactions can involve numerous optical waves within the incident optical field that can have different polarizations. If the material exhibits anisotropic response, meaning that its susceptibility takes different values depending on the orientation of the optical field vector $\mathbf{E}$ with respect to its internal structure, a tensor quantity is assigned to the optical susceptibility to describe its dependence on the orientation of the optical field. This situation typically arises in crystalline solids that commonly exhibit anisotropy because of the directionality of their crystalline structures (see Figs. 2.4a and 2.8b). Since optical crystals represent an essential class of materials in nonlinear optics, we provided Eq. (3.41) for the nonlinear polarization in the most general form to ensure the clarity of the big picture presentation.

Equation (3.41) contains new terms that deserve a discussion. Let us simplify it by assuming the optical fields and polarizations to be scalar quantities. Strictly speaking, this assumption is generally invalid because of the inherently vectorial nature of optical fields, but can still be used for simplicity in illustrating crucial aspects of nonlinear optical interactions. It has limited applicability in situations where all the fields and polarization components are polarized in the same direction. In the case of scalar fields, the susceptibility tensor reduces to a scalar, and Eq. (3.41) becomes

$$P = \epsilon_0 \chi E = \epsilon_0 \chi^{(1)} E + \epsilon_0 \chi^{(2)} E^2 + \epsilon_0 \chi^{(3)} E^3 + \cdots . \tag{3.42}$$

Here, we have optical susceptibility χ that is now a more sophisticated physical quantity containing the information about linear and nonlinear optical responses of the material. By expanding the polarization into the power series with respect to the applied optical field strength, we retrieve separate parts of the susceptibility responsible for the linear and the particular order of nonlinear optical interactions. Specifically, $\chi^{(1)}$ represents linear optical susceptibility that we introduced in Section 3.1. This characteristic describes absorption and dispersion that are linear optical effects. Also, we have the terms $\epsilon_0 \chi^{(2)} E^2$ and $\epsilon_0 \chi^{(3)} E^3$, containing nonlinear optical susceptibilities $\chi^{(2)}$ and $\chi^{(3)}$, responsible for describing nonlinear optical effects of the second- and third-order, respectively. The order of the nonlinear optical effect indicates the power of the strength of the applied optical field with which the corresponding contribution to the polarization scales. In such a way, the contribution $\epsilon_0 \chi^{(2)} E^2$ to the nonlinear polarization scales as the second power of the applied optical field and is responsible for the second-order nonlinear optical interactions, which involve two optical waves from the incident optical beam. The contribution $\epsilon_0 \chi^{(3)} E^3$ to the nonlinear polarization scales as the third power of the applied optical field and is thus responsible for the third-order nonlinear optical interactions, requiring three waves from the incident optical field.

It is generally true that every higher order in the power series expansion of the polarization with respect to the incident field strength is weaker than the lower-order terms. It thus requires stronger optical fields to be observed. The power series expansion (3.42) can incorporate higher-order terms that describe optical nonlinearities where the corresponding contributions to the polarization scale as fourth and higher power of the applied field strength. However, such terms are typically very weak, and the corresponding nonlinear optical phenomena require rather exotic conditions for their experimental observations. In most situations, it is practical to consider the expansion (3.42) up to the third-order nonlinearity because the second- and third-order nonlinear optical effects are readily observable under moderate strengths of the applied optical field. Moreover, these two nonlinearity orders represent an important class of phenomena one needs to be aware of when working with laser light. That is why we limit our consideration to these two contributions to the nonlinear polarization.

For further analysis, let us rewrite Eq. (3.42) in the form

$$P = P^{(1)} + P^{(2)} + P^{(3)} + \cdots , \tag{3.43}$$

where

$$P^{(1)} = \epsilon_0 \chi^{(1)} E, \tag{3.44a}$$

$$P^{(2)} = \epsilon_0 \chi^{(2)} E^2, \tag{3.44b}$$

and

$$P^{(3)} = \epsilon_0 \chi^{(3)} E^3. \tag{3.44c}$$

The term $P^{(1)}$ has the meaning of the linear contribution to the total polarization induced in the medium by the applied optical field. In contrast, $P^{(2)}$ and $P^{(3)}$ represent the second- and third-order nonlinear contributions to the polarization, respectively. Further, in this section, we discuss some specific second- and third-order nonlinear optical phenomena arising from contributions $P^{(2)}$ and $P^{(3)}$ to the polarization.

3.3.2   SECOND-ORDER NONLINEAR OPTICAL PHENOMENA

Second-order nonlinearity is the lowest-order nonlinear optical contribution to the polarization related to the strongest nonlinear contribution $\chi^{(2)}$ to the overall susceptibility. It is an even-order optical nonlinearity, and not every optical material has it as its inherent property because of symmetry considerations that will be discussed later in this section. If present in a given material medium, second-order nonlinear optical effects typically require lower optical power to be observed. The corresponding nonlinear polarization is given by Eq. (3.44b) and can serve as a source to numerous nonlinear optical phenomena of second order. Here, we will briefly overview the most common effects of this kind. For simplicity, let us assume that all the fields and polarization are oriented the same way, which will allow us to drop vector notation.

Second-Harmonic Generation

Second-harmonic generation, associated with doubling the optical frequency of light as it propagates through a nonlinear optical medium, has been described as an example of nonlinear optical interactions at the beginning of this section (see Fig. 3.9). Here, we provide a simplified description for the contribution to the polarization responsible for SHG and explain this phenomenon's essence.

Let us consider a monochromatic optical wave that enters a nonlinear optical medium exhibiting second-order optical nonlinearity, characterized by the contribution to the optical response $\chi^{(2)}$. It is very common to describe the electric field by the function

$$E(t) = E_0 e^{i\omega t} + \text{c. c.,} \tag{3.45}$$

where E_0 is the complex field amplitude and c. c. stands for the complex conjugate. The expanded form of Eq. (3.45) is $E(t) = E_0 e^{i\omega t} + E_0^* e^{-i\omega t}$. Substituting Eq. (3.45) into (3.44b), and separating the components corresponding to the frequency 2ω, one can obtain the doubled-frequency contribution to the polarization of the medium[5]:

$$P^{(2)}(2\omega) = \epsilon_0 \chi^{(2)}(2\omega; \omega, \omega) E_0^2 e^{i2\omega t} + \text{c. c..} \tag{3.46}$$

This second-harmonic component in the spectrum of oscillation of the optical polarization serves as a source to a secondary optical wave re-emitted by the nonlinear optical medium at frequency 2ω. This frequency doubling effect is never 100% efficient, i.e., not all the fundamental radiation gets converted into the second harmonic. As a result, at the output of the medium, we have the residual fundamental radiation and the new frequency component originating from the SHG process. In Fig. 3.10a, we provide a cartoon illustration of this process.

Like any other physical effect, SHG requires the fundamental laws of energy conservation and momentum conservation to be fulfilled. The energy conservation law can be schematically represented on a diagram shown in Fig. 3.11a. The mathematical equivalent of this representation can be expressed as

$$\hbar\omega + \hbar\omega = 2\hbar\omega, \tag{3.47}$$

where $\hbar\omega$ is the energy of a fundamental photon while $2\hbar\omega$ is the energy of a second-harmonic photon. In such a way, two fundamental photons are consumed in the nonlinear optical inter-

---

[5]The substitution will also reveal that mixing the terms proportional to $\exp(i\omega t)$ and $\exp(-i\omega t)$ results in zero-frequency contribution to the polarization. This contribution describes the nonlinear optical phenomenon called *optical rectification*. For more information about this and other nonlinear optical effects not considered in this section, we refer the reader to more specialized books on nonlinear optics [19, 21, 22].

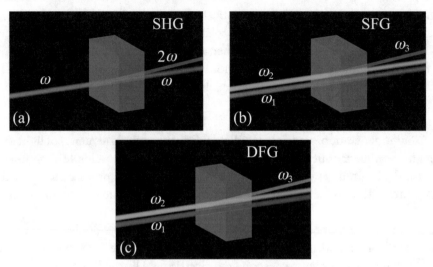

Figure 3.10: Schematic representation of some second-order nonlinear optical effects. (a) Second-harmonic generation: the fundamental beam of frequency ω gets partially converted into a doubled-frequency beam 2ω in a nonlinear optical crystal. (b) Sum-frequency generation: two incident beams with frequencies ω_1 and ω_2 mix together in a nonlinear optical crystal to produce a new spectral component with frequency $\omega_3 = \omega_1 + \omega_2$. (c) Difference-frequency generation: two incident beams with frequencies ω_1 and ω_2 mix together in a nonlinear optical crystal to produce a new spectral component with frequency $\omega_3 = \omega_1 - \omega_2$.

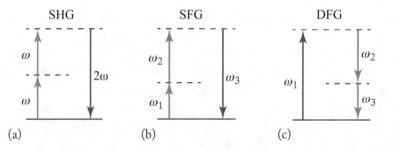

Figure 3.11: Energy diagrams of some second-order nonlinear optical effects. (a) Second-harmonic generation: two fundamental photons of frequency ω get consumed by a nonlinear optical process that produces a doubled-frequency photon of frequency 2ω. (b) Sum-frequency generation: two incident photons with frequencies ω_1 and ω_2 get consumed by a nonlinear optical process that produces a new photon with frequency $\omega_3 = \omega_1 + \omega_2$. (c) Difference-frequency generation: two incident photons with frequencies ω_1 and ω_2 get consumed by a nonlinear optical process that produces a photon with frequency $\omega_3 = \omega_1 - \omega_2$. Solid lines indicate real (existing) energy states, while dashed lines represent virtual (non-existing) energy states.

action process to produce a new photon at doubled frequency. The diagram shown in Fig. 3.11a indicates that SHG is a *parametric* nonlinear optical process where one starts at a lower atomic energy level and ends up at the same level after each act of an SHG photon generation. Therefore, the energy conservation law is automatically fulfilled in this process. However, the momentum conservation law is not as straightforward, which we will discuss in the following subsection.

There is one crucial point that concerns the energy diagrams represented in Fig. 3.11. The real energy states which exist in a given atomic or molecular system are shown with solid lines. There are some intermediate states involved in the nonlinear optical processes shown in the diagrams on the figure represented by the dashed lines. These states are virtual; they do not exist in the atomic system in reality. They are marked at the diagram to guide the reader on the relative energies of the photons involved in the nonlinear optical process. Remarkably, the upper energy level of all these diagrams is virtual. It means that the nonlinear optical process cannot terminate there, and the final energy state of the process becomes the ground state, the only real state in the process. This ensures that the energy conservation law is satisfied within the nonlinear optical process itself without resorting to other channels of energy dissipation, unlike in nonlinear absorption that we will discuss at the very end of the chapter (see Fig. 3.21).

Phase-Matching Condition

The momentum conservation law requires that the wave vectors of the interacting fundamental and second-harmonic waves satisfy the condition

$$\mathbf{k}(\omega) + \mathbf{k}(\omega) = \mathbf{k}(2\omega). \tag{3.48}$$

Equation (1.11) translates this into *phase-matching condition*: the phases of the fundamental and second-harmonic waves need to be matched at every point in space so that there is no walk-off between the waves, and the nonlinear optical interaction occurs with maximum efficiency. Lack of phase matching between the interacting waves can result in loss of efficiency of the non-linear optical process. We illustrate this concept in Fig. 3.12 where we schematically show the nonlinear optical medium (blue rectangle), the fundamental (red lines), and second-harmonic (blue lines) waves. In an ideal scenario of perfect phase matching between the fundamental and second-harmonic waves, depicted in Fig. 3.12a, the fundamental wave propagates through the nonlinear optical medium and gradually gives out its energy to the SHG nonlinear process, thereby magnifying the energy of the second-harmonic wave. Part (b) of the figure simplistically illustrates a more realistic scenario where the phases of the fundamental and second-harmonic waves are not matched. The waves experience walk off with respect to each other, and the cor-responding harmonic generation process lacks efficiency.

What does this phase matching entail, and why is it not fulfilled automatically? To an-swer these questions, let us consider the simplest case of a collinear interaction where both the fundamental and second-harmonic waves propagate in the same direction. This arrangement is schematically represented in Fig. 3.13a. The fundamental and second-harmonic waves are shown with red and blue lines, respectively, while their wave vectors are represented with the

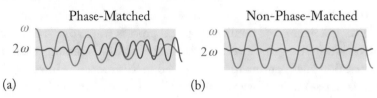

Figure 3.12: Second-harmonic generation: (a) phase-matched scenario, where $2k(\omega) = k(2\omega)$; and (b) non-phase-matched scenario where $2k(\omega) \neq k(2\omega)$.

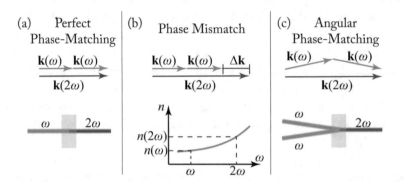

Figure 3.13: Phase-matching diagrams. Part (a) describes a perfect phase-matching. Part (b) explains the origin of phase mismatch. The top portion shows phase mismatch acquired between the fundamental and second-harmonic wave vectors. The bottom portion shows the dispersion of the refractive index, inevitably leading to $n(\omega) \neq n(2\omega)$, and the phase mismatch. (c) Angular phase matching, achievable by selecting a proper angle between the interacting fundamental waves.

corresponding color's arrows. The case of perfect phase matching, given by Eq. (3.48), is illustrated at the top of Fig. 3.13a: the sum of the momenta of two waves from the fundamental beam should be equal to the momentum of the second-harmonic wave. In the case of collinear interaction, Eq. (3.48) evolves into the requirement for the corresponding wavenumbers (absolute values of the wave vectors) to satisfy the relationship

$$2k(\omega) = k(2\omega). \tag{3.49}$$

Let us recall that we are dealing with light propagation in a medium. In agreement with Eq. (1.4), relating the wave vectors in vacuum and in a medium, Eq. (3.49) can be rewritten as

$$2n(\omega)k_0(\omega) = n(2\omega)k_0(2\omega). \tag{3.50}$$

It is evident from Eqs. (1.4), (1.1), and (1.7) that $k_0(\omega) = \omega/c_0$, and the relationship (3.50) can be rewritten as

$$2n(\omega)\frac{\omega}{c_0} = n(2\omega)\frac{2\omega}{c_0}. \qquad (3.51)$$

The condition given by Eq. (3.51) can only be satisfied if

$$n(\omega) = n(2\omega). \qquad (3.52)$$

It means that, in order for the momentum conservation law to be fulfilled, it is necessary for the refractive indices of the medium at the fundamental frequency ω and at the second-harmonic frequency 2ω to be equal. The frequencies ω and 2ω differ by a factor of two, which is a lot from the standpoint of dispersion. Even if both the frequencies fall within a range where the dispersion is low, and the refractive index change with the variation of frequency occurs very slowly, for any natural optical material, there will be considerable difference between $n(\omega)$ and $n(2\omega)$ to make Eq. (3.52) unsatisfiable. This is illustrated in Fig. 3.13b where we show a realistic scenario of non-phase-matched collinear interaction, characterized by a phase mismatch

$$\Delta\mathbf{k} = \mathbf{k}(2\omega) - 2\mathbf{k}(\omega) \neq 0 \qquad (3.53)$$

(the top portion of the figure), and a typical frequency dependence of the refractive index within a window of transparency (the bottom portion). There exist several phase-matching techniques one can follow to achieve the phase-matching condition between the fundamental and second-harmonic waves. One example is displayed in Fig. 3.13c. The interacting fundamental beams fall onto the nonlinear optical medium at an angle to each other (the bottom portion). The top portion of the figure shows how this arrangement modifies the phase-matching condition where the fundamental wave vectors match the SHG wave vector being at the right angle to each other. This approach is termed *angle phase matching* and represents only one out of the number of existing phase-matching techniques. Before we delve into describing these techniques, it is essential to remark that everything said in this and the following subsection is valid for all parametric nonlinear optical effects. We use SHG merely as an example to explain the concept of phase matching, its importance in nonlinear optics and the tricks one can rely on to achieve it.

Phase-Matching Techniques

All parametric nonlinear optical interactions require an appropriate phase-matching condition to be fulfilled. Among the methods used to satisfy condition (3.49) are the following.

Natural Birefringence

One conventional method to achieve phase matching in nonlinear optics is to exploit natural *birefringence* of nonlinear optical crystals. Birefringence is a property of many crystals where the refractive index takes different values in different directions with respect to their crystalline lattices. This property is typical for anisotropic crystals with noncubic crystalline lattices where different crystalline directions have different lattice (unit cell) constants and different atomic arrangements. An optical wave propagating through an anisotropic crystal can thus experience different delays and phase shifts, depending on its propagation direction within the crystal and polarization. Moreover, it can assume two alternative propagation paths and, consecutively, split into two sub-waves within a birefringent crystal. Even in the simplest case of a *uniaxial* birefringent crystal, some propagation directions within such a crystal can simultaneously exhibit two distinct values of the refractive index, corresponding to two different polarization orientations. Uniaxial crystals are characterized by a single direction within the crystal along which there is no birefringence. This direction is called *optic axis* of the crystal. For a *biaxial* crystal, there are two such directions, indicating the presence of two optic axes.

We further consider uniaxial crystals as an example of birefringent materials used for phase-matched SHG. A light beam, propagating through such a crystal along some direction other than its optic axis, can split into two beams: *ordinary* and *extraordinary*. Such splitting occurs because there are two values of the refractive index associated with the crystal, ordinary n_o and extraordinary n_e, which can simultaneously dictate different propagation conditions to the beam.

The names "ordinary" and "extraordinary" originate from the following properties of these refractive index values. Ordinary refractive index n_o does not change its value with the orientation of the polarization of light. In contrast, the value of extraordinary refractive index n_e depends largely on the polarization of the incident light. This property of the two refractive indices is illustrated in Fig. 3.14a where we schematically show n_o and n_e for a negative ($n_e \leq n_o$) uniaxial crystal. As one can see from the figure, there is a single direction, lying along the optic axis of the crystal, in which $n_o = n_e$. In all other directions, $n_e < n_o$. The parameter Δn, equal to the maximum difference between n_o and n_e, achievable at a specific wavelength, characterizes the strength of the crystal's birefringence.

Finally, one can draw a diagram similar to the one shown in Fig. 3.14a for a specific wavelength. It will have a different size for a different wavelength since the refractive indices have different values at different wavelengths.

Let us consider a specific example of a negative uniaxial nonlinear optical crystal, β-BaB$_2$O$_4$ (BBO), which is frequently used for SHG in a wide range of wavelengths falling within its transparency window 190–3,300 nm. The fundamental wavelength falling in the range between 410 and 3,500 nm can be efficiently converted to a second harmonic in BBO. In Fig. 3.14b, we show n_o and n_e refractive indices of BBO as functions of wavelength for a beam having the propagation direction and polarization orientation such that the birefringence of the crystal is maximal. The value of n_o does not depend on the angle of incidence. The value of n_e, on the contrary, changes with the angle of incidence in the range between the blue line (minimal value of n_e) and solid black line, representing n_o and corresponding to the maximal value of n_e achievable along the optic axis (when both refractive indices are equal). These curves demonstrate the range within which one can tune the refractive indices of the fundamental and

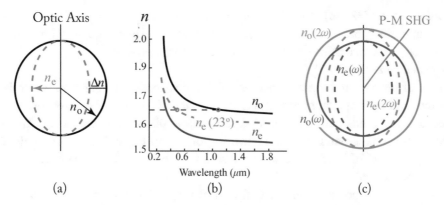

Figure 3.14: (a) Ordinary (circle) and extraordinary (ellipse) refractive indices of a negative uniaxial nonlinear optical crystal. The birefringence is schematically shown by the difference Δn between the values of the refractive indices. (b) Ordinary and extraordinary refractive indices of β-BaB$_2$O$_4$ (BBO) crystal as a function of wavelength. (c) An example of phase-matched SHG (P-M SHG) in a uniaxial nonlinear optical crystal. The brown (green) circle and ellipse represent the n_o and n_e for the fundamental (second-harmonic) wave, respectively. The red line shows the condition at which P-M SHG can occur.

second-harmonic beams to find the conditions under which they are equal. How, in particular, one can achieve such a condition?

In Fig. 3.14c, we show a schematic representation of phase-matched SHG in a crystal similar to BBO. There are two sets of circles and ellipses at the picture, showing n_o and n_e refractive indices at two different frequencies of light: the fundamental frequency ω (brown curves) and the second-harmonic frequency 2ω (green curves). One can see that the circle and ellipse signifying the refractive indices at the fundamental frequency are smaller compared to those for the second harmonic because of normal material dispersion (see Fig. 3.14b showing the growth of the refractive indices with an increase of the frequency of light). By a careful examination of part (c) of the figure, one can also remark that there is a direction in which $n_o(\omega) = n_e(2\omega)$; it is shown with the solid red line. This direction corresponds to phase-matched SHG. One can achieve this condition by a proper selection of the polarization of the fundamental wave and its angle of incidence with respect to the optic axis of the nonlinear optical crystal. In particular, as an example, one can convert 1,064-nm IR radiation into 532-nm visible light (green), achieving the equality of the refractive indices $n_o(1,064 \text{ nm}) = n_e(532 \text{ nm}) \approx 1.65$ in BBO crystal by properly selecting the angle of incidence of the fundamental beam with respect to the optic axis (around 23 deg). This example is represented in Fig. 3.14b with red dashed lines, showing the refractive index n_e in the case when the incident fundamental beam makes around 23° with respect to the optic axis of BBO. One can see from the graph by observing black solid and red dashed lines that $n_e(532\,\text{nm}) = n_o(1,064\,\text{nm})$, which makes phase matching for this experimental configuration possible. For more information about exploiting natural birefringence of the crystals for phase-matching we refer the reader to the source [23].

Angular Phase-Matching

This method of phase matching is associated with the natural birefringence, which is a property of some nonlinear optical crystals, as described above. By a proper selection of the angle of incidence of the fundamental waves, one can achieve phase-matched SHG. An additional illustration of this phase matching mechanism is provided in Fig. 3.13c, where we demonstrate how the proper angles of incidence resolve the situation shown in Part (b) of the figure.

Dispersion Engineering

Bulk optical materials possess their natural optical properties, including material dispersion, which cannot be modified and can only be worked around in some cases (e.g., exploiting natural birefringence in some nonlinear optical materials). *Dispersion engineering* refers to tailoring dispersion in optical materials through micro and nanostructuring or through creating artificial structures out of bulk materials. Some examples of such structures are *optical fibers* and *optical waveguides*, which will be discussed later in the book series.

Some more exotic examples include *photonic crystal structures*, representing periodic structures with significantly modified optical properties compared to their bulk constituents, as well as nanopatterned materials. For the latter, one has to rely on nanofabrication technologies, which are available nowadays. There is an abundance of interesting physical effects and methods of controlling the flow of light enabled by this approach. Enhanced phase-matched nonlinear optical interactions are only one side of what such artificial materials can offer. This topic is, however, very advanced and is outside the scope of this book. For more information, we refer the reader to research articles [24–26] that summarize recent achievements in nanostructured optical materials.

An artificial structure, such as optical fiber, waveguide, or nanostructured optical device, has a complex geometry involving numerous optical media among which the propagating light distributes itself. Since the light "sees" a sophisticated structure with different parts exhibiting different optical properties, the effective (average) refractive index that it experiences can differ significantly from the material refractive index of a bulk, non-structured optical medium. This difference is wavelength-dependent, which can serve as a means for achieving the equality of the effective refractive indices for different wavelengths associated with a nonlinear optical process.

The phase-matching approaches described in this section are readily applicable to the whole class of parametric nonlinear optical processes apart from SHG used as an example in this discussion. We further proceed to describe some other second-order nonlinear optical effects.

Sum-Frequency Generation

Sum-frequency generation (SFG) represents a nonlinear optical effect similar to SHG, with the difference that there are two distinct spectral components (colors of light) at the entrance to the nonlinear optical medium, characterized by the optical fields E_1 and E_2 with the frequencies ω_1 and ω_2, respectively. The fields satisfy the same mathematical form as Eq. (3.45):

$$E_i(t) = E_{0i}e^{i\omega_i t} + \text{c. c..} \tag{3.54}$$

Here, the index $i = 1, 2$ represents wave E_1 or E_2, and E_{0i} is the amplitude of the corresponding wave.

One of the scenarios for the waves' nonlinear optical interaction in a material exhibiting $\chi^{(2)}$ is

$$\hbar\omega_1 + \hbar\omega_2 = \hbar\omega_3, \tag{3.55}$$

so that a new spectral component with the frequency equal to the sum of the frequencies of the incident spectral components is generated. The process is illustrated in Fig. 3.10b. The energy diagram of the process is shown in Fig. 3.11b, and the corresponding contribution to the nonlinear polarization can be described by equation

$$P^{(2)}(\omega_1 + \omega_2) = 2\epsilon_0 \chi^{(2)}(\omega_1 + \omega_2; \omega_1, \omega_2)E_1 E_2. \tag{3.56}$$

Everything that we mentioned in relevance to SHG applies to SFG. Like SHG, it is a parametric nonlinear optical effect and requires a phase-matching condition to be satisfied. One can use similar approaches described in the context of SHG to achieve phase-matched SFG.

Difference-Frequency Generation

Let us assume that we have an incident optical field carrying two spectral components with the frequencies ω_1 and ω_2, each of which satisfying Eq. (3.54). Let us now pay attention to the fact that Eq. (3.54) for the incident optical field components contains complex conjugate terms proportional to $\exp(-i\omega_i t)$. It implies that optical frequencies cannot only be added up to produce SFG but can also be subtracted from each other to produce *difference-frequency generation* (DFG). A picture, shown in Fig. 3.10c, illustrates this process. The corresponding energy diagram, describing the frequency mixing process $\omega_1 - \omega_2 = \omega_3$, is shown in Fig. 3.11c. The corresponding contribution to the nonlinear polarization can be expressed as

$$P^{(2)}(\omega_1 - \omega_2) = 2\epsilon_0 \chi^{(2)}(\omega_1 - \omega_2; \omega_1, -\omega_2)E_1 E_2^*, \tag{3.57}$$

where E_2^* is the complex conjugate of the second incident wave. The equivalent of energy conservation law for this process has the form

$$\hbar\omega_1 - \hbar\omega_2 = \hbar\omega_3. \tag{3.58}$$

Like SHG and SFG, the DFG process holds promise where one requires access to light with the frequencies that cannot be easily generated using conventional light sources. The difference between these processes is that DFG allows one to access the range of longer wavelengths, such as near and mid-IR. An example of an optical device that produces the radiation with the

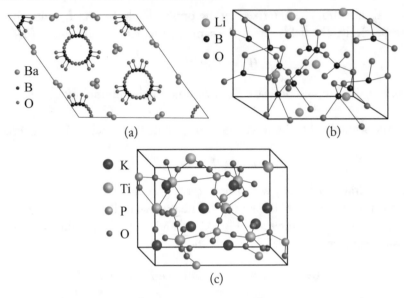

Figure 3.15: Crystalline structures of some non-centrosymmetric crystals conventionally used for SHG and for generating other second-order nonlinear optical effects: (a) BaB_2O_4 (BBO), (b) LiB_3O_5 (LBO), and (c) KO_5PTi (KTP). Parts (a) and (b) were reproduced in similarity to Fig. 3 from [27]; and Part (c) was reproduced in similarity to Fig. 1 from [28].

wavelength spanning the range between 1 and 4 μm is an optical parametric oscillator (OPO). Its operation is based on the DFG effect that proves extremely useful for this application. In contrast, SHG and SFG target the shorter-wavelength range by doubling and summing up the frequencies of the incident waves.

Centrosymmetric and Non-Centrosymmetric Optical Materials

In order for a material to exhibit nonlinearity of the second (and higher even) orders, its structural element, such as the atomic unit cell, has to lack the center of symmetry. Such materials are called *non-centrosymmetric*. What exactly do we mean by the absence of the center of symmetry? When looking from the center of a unit cell of a non-centrosymmetric crystal, one finds that its opposite points situated at the coordinates x, y, z and $-x$, $-y$, $-z$ to the center are different. Some examples of non-centrosymmetric materials exhibiting even-order nonlinearities are nonlinear optical crystals such as BaB_2O_4 (BBO), LiB_3O_5 (LBO), and KO_5PTi (KTP). Their structures are depicted in Fig. 3.15.

In case of a centrosymmetric crystal, such as, for example, MgO (see Fig. 2.8b), any point with the coordinates x, y, z is indistinguishable from the point with the coordinates $-x$, $-y$, $-z$. Such crystals are said to exhibit inversion symmetry. Moreover, amorphous materials do not have a definite structure and therefore lack the properties of non-centrosymmetric materials. They thus cannot exhibit even-order nonlinear optical effects. That is why in order to observe second-order nonlinear optical processes in a bulk material, one has to select a non-centrosymmetric crystal.

3.3.3   THIRD-ORDER NONLINEAR OPTICAL PHENOMENA

Third-order nonlinear optical effects arise from the contribution to the overall polarization induced by the incident optical field, given by Eq. (3.44c). In contrast to even-order nonlinearities, odd-order nonlinearities are inherent to any material, whether amorphous or crystalline, centrosymmetric, or non-centrosymmetric. Every optical material is nonlinear, with some materials exhibiting a stronger optical nonlinearity while others having much weaker nonlinear optical responses. The materials with strong nonlinear optical responses, characterized by higher nonlinear susceptibilities, require less optical intensity in order for the nonlinear interactions to be experimentally observable. Below we outline some common third-order nonlinear optical phenomena. Some of them, such as third-harmonic generation (THG), are similar to those occurring in the second-order nonlinear optical responses.

In contrast, some other ones, such as the intensity-dependent refractive index, are typical for odd-order nonlinear optical responses. Even though a second-order nonlinear optical response is typically much stronger, third-order optical nonlinearities are still quite commonly observable under typical experimental conditions. They can be useful or parasitic, depending on the specific case. Moreover, in amorphous and centrosymmetric materials, third-order nonlinearities represent the lowest-order nonlinear optical interactions. That is why it is important to be aware of these effects.

Some Third-Order Nonlinear Effects

Some third-order nonlinear optical effects are somewhat similar to those discussed above in the context of second-order nonlinearities. A classic example is a THG, which is similar to SHG, except that there are three fundamental photons consumed by the nonlinear optical process to produce a tripled-frequency photon. The corresponding photon energy conservation law requires that

$$\hbar\omega + \hbar\omega + \hbar\omega = 3\hbar\omega. \tag{3.59}$$

THG is a parametric nonlinear optical effect that can be described by the following contribution to the overall polarization:

$$P^{(3)}(\omega + \omega + \omega) = \epsilon_0 \chi^{(3)}(3\omega; \omega, \omega, \omega)E^3. \tag{3.60}$$

All the considerations such as phase-matching conditions and requirements of sufficient optical intensity for the effect to be significant apply to all the parametric third-order nonlinear optical processes. Since THG converts the frequency of light to a three-times higher frequency, it is beneficial for producing radiation at the wavelength ranges where the conventional practical light sources do not operate.

In addition to THG, there exist the third-order SFG and DFG, involving three incident waves that can be of different frequencies. Since they are quite similar to the second-order SFG and DFG, we do not describe them here. More information about these effects is available in the *Nonlinear Optics* book by R. W. Boyd [19]. Below we describe some nonlinear optical phenomena unique to odd-order nonlinear optical processes and not discussed in the context of second-order nonlinearities.

Nonlinear Refractive Index

Nonlinear contribution to the refractive index, also called *Kerr effect*, is a third-order nonlinear optical process where light propagating through a nonlinear optical medium modifies the medium's material response proportionately to its intensity. In other words, the light changes propagation condition of its own by inducing an intensity-dependent change to the medium's refractive index. The overall refractive index $\tilde{n}$ of the medium, now including both linear and intensity-dependent contributions, is described by the equation

$$\tilde{n} = n + n_2 I. \tag{3.61}$$

Here, the intensity-dependent contribution is $n_2 I$, where n_2 is the *Kerr coefficient*, one of the nonlinear material response parameters. The value and sign ($\pm$) of n_2 are unique to a specific material. When either the value of n_2 of a given optical material or the light's intensity I is sufficiently low, such that $n_2 I$ is negligibly small, the intensity-dependent contribution to the refractive index can be neglected. In this case, we are dealing with linear light propagation through an optical medium, characterized by the linear refractive index n, as described earlier in the book. In most of the cases, this is valid, and we can concentrate on describing the accompanying linear optical effects, such as absorption and dispersion. However, it is not always the case, and one needs to be aware of the impact of the nonlinear refractive index on light propagation.

There exist numerous manifestations of such an intensity-dependent change to the refractive index. The most common accompanying phenomenon is *self-focusing* (or *self-defocusing*, depending on the sign of n_2) of an optical beam. The intensity-dependent contribution to the refractive index of the material introduces an intensity-dependent change in the propagating beam's phase. This nonlinear phase contribution can vary from point to point as the intensity of the beam changes. As a consequence, the conditions of the beam's propagation change along its propagation direction. It can result in the modification of the beam's wavefront's curvature, causing the beam's focusing or defocusing. Specifically, a plane wave representing an originally collimated beam of light can evolve into a converging or diverging spherical wave, experiencing focusing or defocusing, as demonstrated in Fig. 3.16a and b, respectively.

Self-focusing and self-defocusing are frequently recognized as parasitic nonlinear optical effects. On the other hand, they can be useful if implemented in a controlled manner. Thus, it

Self-Focusing

Self-Defocusing

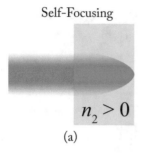

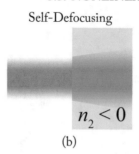

$n_2 > 0$

$n_2 < 0$

(a)

(b)

Figure 3.16: (a) Self-focusing of a beam propagating through an optical medium with $n_2 > 0$. (b) Self-defocusing of a beam propagating through a medium with $n_2 < 0$.

is important to be aware of these nonlinear optical effects to be able to mitigate or implement them where necessary.

Let us consider, for example, self-focusing. The light propagating through a nonlinear optical medium with $n_2 > 0$ undergoes self-focusing and can focus on a very small volume. All the optical power carried by the light beam thus gets concentrated into that small volume. In some instances, such a large amount of power in a small volume can cause optical damage to the material medium through which the light propagates. Optical damage is irreversible and creates defects, scattering, and absorption centra in an initially uniform and lossless material, which makes the material not suitable for further use in optical setups. This situation is frequently encountered in laser crystals, optical components such as polarizers, wave plates, filters, etc. that represent critical parts of laser resonators. In this context, self-focusing plays the role of a parasitic optical effect, which requires consideration and ways of mitigation.

Let us consider a different situation where self-focusing is implemented with care to achieve a useful result. In pulsed laser systems, one can use self-focusing to achieve *mode-locking* necessary for producing ultrashort (with picosecond and femtosecond duration) pulses of laser radiation.

Self-focusing and self-defocusing belong to a group of nonlinear optical phenomena named *self-action effects*. The name of this group speaks for itself: the light propagating through a nonlinear optical medium modifies the propagation conditions of its own (acts on itself). The corresponding nonlinear contribution to the medium's overall polarization, describing self-action effects associated with the third-order nonlinearity, is represented by

$$P^{(3)}(\omega) = 3\epsilon_0 \chi^{(3)}(\omega; \omega, -\omega, \omega)|E|^2 E. \tag{3.62}$$

It is clear from Eq. (3.62) that the optical field nonlinearly interacts with itself. As an outcome of such interaction, the beam induces a contribution to the refractive index, proportional to its intensity ($\propto |E|^2$).

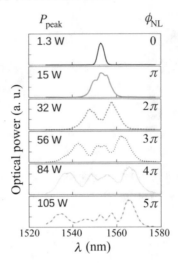

Figure 3.17: Spectral modifications encountered by an optical pulse as it propagated through a semiconductor nonlinear optical waveguide structure. The spectral modifications evolved with the change in the optical power at the entrance to the waveguide. The pulse's peak power, P_{peak}, is shown with labels on the left-hand side of the graphs, and the corresponding acquired nonlinear phase shift, ϕ_{NL}, is shown on the right-hand side of the graphs. $P_{\text{peak}} = 1.3$ W, corresponding to $\phi_{\text{NL}} = 0$, matches the initial spectrum of the pulse as originated from the laser. The figure was extracted from the author's real-life experiment on SPM in an AlGaAs waveguide.

Another notable representative of the group of self-action effects is *self-phase modulation* (SPM). This effect is frequently encountered in optical fibers where the propagation distances are very long, resulting in extensive lengths of the nonlinear optical interaction of light with the medium of the fiber. Like other self-action effects, SPM is associated with the nonlinear contribution to the refractive index n_2 (Kerr effect) and the nonlinear phase shift induced by this contribution. The manifestation of the nonlinear phase shift acquired by an optical beam propagating through a nonlinear optical medium for a sufficiently long distance is the beam's spectral broadening or the appearance of new spectral components in the beam's output spectrum. The optical beam's original spectrum can become modified significantly in the nonlinear optical medium if the interaction length or the strength of the nonlinear optical coefficients is high.

In Fig. 3.17, we show an example of spectral modifications experienced by an optical pulse due to SPM, derived from a real-life experiment. Optical pulses were allowed to propagate through a semiconductor optical waveguide, exhibiting a very strong optical nonlinearity (a very high value of $n_2 = 1.5 \times 10^{-13}$ cm²/W). The pulses' initial strength was evaluated in terms of their pick power P_{peak}, which varied between 1.3 W and 105 W in the experiment. The power values are displayed on the labels on the left-hand sides of the graphs. When the pulses' peak power at the entrance to the waveguide was 1.3 W, it was insufficient to observe any impact of SPM on the pulse spectrum. The resulting spectrum, shown at the very top of the figure (power in arbitrary units as a function of the wavelength of light), matches the original spectrum of the pulse emitted by the laser. As a result, the nonlinear optical phase shift φ_{NL}, acquired by the pulse, was zero for this lowest power setting. In this case, the optical pulse propagating through the waveguide structure interacted with the material medium linearly; there was no noticeable nonlinear optical interaction involved. As the power was gradually increased (top to bottom), the nonlinear optical interaction started to develop. The nonlinear phase shift φ_{NL} gained higher and higher values, as shown with labels on each graph's right-hand side. The corresponding spectrum of the pulse at the waveguide's output exhibited significant modifications. Specifically, it acquired more and more peaks and became broader with the pulse power increase. In different words, the spectrum of the pulse acquired new spectral components, not present in the original spectrum, through the nonlinear optical broadening.

Similar to self-focusing, SPM can be regarded as both a parasitic and a beneficial effect. Its parasitic side is that, in many situations, spectral modifications to optical pulses are highly undesirable. Apart from the fact that one loses control over the spectral shape of the pulse broadened by SPM, the pulse acquires chirp and spreads in time, becoming non-transform-limited. The temporal spread of optical pulses can, for example, adversely affect information transmission in optical communication networks, limiting their speed of operation.

Among the useful sides of SPM, there is the ability to obtain new spectral components non-existing in an original optical pulse. If we take SPM to an extreme (by producing highly favorable conditions for strong nonlinear optical interaction), the pulse's spectrum can evolve into a *super-continuum* spectrum, covering an enormous wavelength band. As a specific example, in a specialty fiber designed explicitly for nonlinear optical interactions, it is possible to generate all colors of the rainbow within the visible spectral window out of an IR seeding optical pulse! Having an IR at the input, we can thus obtain a rainbow of colors at the output, which can prove useful for myriad applications, such as spectroscopy, imaging, metrology, and many more.

An example of supercontinuum source radiation is shown in Fig. 3.18a. The beam of white light, generated by seeding IR radiation in a nonlinear optical fiber, is dispersed by a grating, which allows exposing its various spectral components. In Fig. 3.18b, we also show a spectrum of a supercontinuum source. The original spectrum of IR radiation, which initiates supercontinuum generation via interacting with the optical fiber nonlinearly, is shown with solid blue lines, and the supercontinuum spectrum is shown with solid red lines. One can see from the graph that the original pulse has the peak wavelength at around 1.07 μm, and its spectrum is narrow. In contrast, the supercontinuum spectrum spans from 0.4 to over 1.7 μm, covering the entire visible window and beyond.

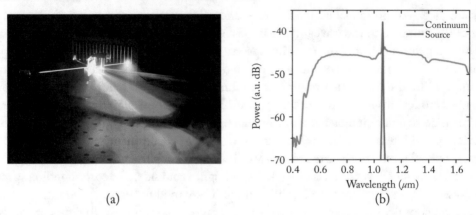

(a) (b)

Figure 3.18: (a) Experimental demonstration of supercontinuum generation through SPM in a nonlinear optical fiber. The supercontinuum output appears as a white beam of light containing all the visible spectrum components and beyond. The beam is dispersed by a grating, which allows exposing the variety of colors incorporated within its spectrum. The image is a property and courtesy of NKT Photonics (https://www.nktphotonics.com/lasers-fibers/technology/supercontinuum/). (b) An example spectrum of a supercontinuum source. The original infrared spectrum of the optical radiation initiated SPM process is shown with the solid blue lines. The supercontinuum spectrum is shown with the solid red lines. ©User:Burlyc / Wikimedia Commons / CC BY-SA 3.0 (https://en.wikipedia.org/wiki/Supercontinuum).

Four-Wave Mixing

Four-wave mixing (FWM) is a widespread third-order nonlinear optical process that can frequently be encountered in various practical situations. Like other nonlinear optical effects, it can prove extremely useful in some situations, while it can be regarded as parasitic in other situations. We find it essential to say a few words about this effect in our summary of nonlinear optics.

FWM can be realized in the following two scenarios. The first scenario is the most general case where all interacting beams of light have different frequencies. The energy diagram is presented in Fig. 3.19a, and the cartoon illustration of the process is shown in Fig. 3.19c. If there are three optical beams with different frequencies at the entrance to the nonlinear optical medium, they can interact via the FWM process to produce the new, fourth-frequency component. The corresponding equivalent of the energy conservation law is given by the equation

$$\hbar\omega_{p1} + \hbar\omega_{p2} - \hbar\omega_s = \hbar\omega_i. \tag{3.63}$$

Here, ω_{p1} and ω_{p2} are the frequencies of the two optical beams at the entrance to the medium, called "pump 1" and "pump 2," ω_s is the frequency of the third spectral component at the input to the medium, called the "signal," and ω_i is the frequency of the newly generated component,

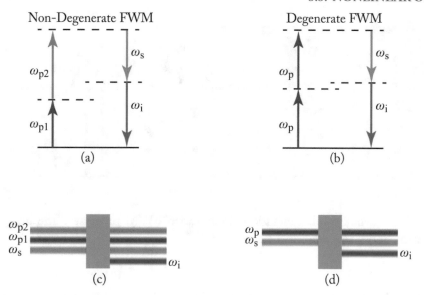

Figure 3.19: Energy level diagrams for (a) non-degenerate and (b) degenerate four-wave mixing processes. Cartoon illustrations of (c) non-degenerate and (d) degenerate four-wave mixing processes.

called the "idler." The terminology comes from the fact that the pump beams, generally carrying more optical power, give out their energies to amplify the weaker signal beam and produce the new frequency component called the idler.

The second scenario for the realization of the FWM process involves only one pump beam and only two distinct spectral components at the entrance to the nonlinear optical medium. Its energy level diagram and cartoon illustration are shown in Fig. 3.19b and d, respectively. The corresponding energy conservation law is described by the equation

$$2\hbar\omega_p - \hbar\omega_s = \hbar\omega_i : \tag{3.64}$$

two pump photons and one signal photon get consumed in the nonlinear optical process to produce a new idler photon with the frequency ω_i. Alternatively, one can view this process as the consumption of two pump photons to generate a signal and an idler photon where the signal beam stimulates the process to occur for a specific signal-idler frequency combination. Because there is a single pump beam involved, this FWM process is called *degenerate*.

FWM is a parametric nonlinear optical process that requires phase matching. There are many frequency components involved in the process, and it is seemingly difficult to achieve phase matching. On the other hand, if one thinks of harmonic generation processes, such as SHG and THG considered herein, there we are dealing with a giant difference in frequency between the fundamental and harmonic radiation: it is two-fold for SHG, and three-fold for

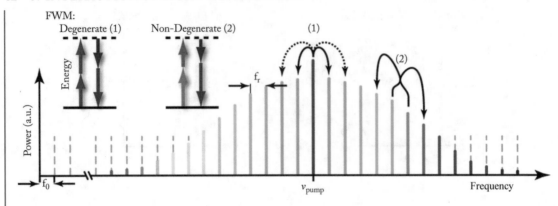

Figure 3.20: Frequency comb generation via cascaded FWM processes. The image is in the public domain, and it was reproduced from Ref. [29].

THG. In FWM, the frequency components can be very close to each other, such that the material dispersion does not represent such a big concern for the phase matching. As a result, FWM is relatively easy to observe. In some situations, it accompanies the flow of existing physical processes in a way that one cannot eliminate this effect easily. Because FWM leads to spectral modifications of the propagating radiation (and sometimes can be accompanied by other nonlinear effects such as SPM), it is generally considered a parasitic effect. This is especially true in optical communication networks where FWM occurring in optical fibers can severely distort the information carried by the light streams. Moreover, it can lead to crosstalk between the communication channels transmitted by an optical fiber.

An example of a useful aspect of FWM is parametric amplification. The fact that the more powerful pump frequency components yield their energy for producing idler and amplifying signal is in the basis of this application. Another example is frequency comb generation, where the cascaded FWM process is implemented to produce regular frequency peaks over a wide spectral window. This process can start from two frequency components at the input to the nonlinear optical medium. If conditions for enhanced FWM are created, the two input frequency components can give rise to an idler component and interact with the newly generated idler to produce more frequency peaks. This effect requires a very high optical nonlinearity, dispersion management, an optical resonator that selectively resonates with specific frequencies coinciding with the ones involved in the FWM process. This process can continue in a cascaded manner if the interaction length is sufficiently high, which is achievable in a high-quality optical resonator. The outcome of such a resonant cascaded FWM process is an entire set of frequency peaks, regularly spaced, that can occupy a very wide spectral window (see Fig. 3.20).

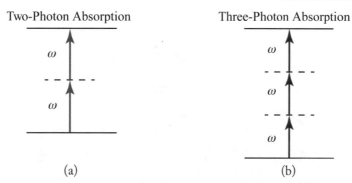

Figure 3.21: Energy diagram of (a) two-photon absorption and (b) three-photon absorption processes.

Nonparametric Nonlinear Optical Processes

So far, we have speculated about parametric nonlinear optical processes where the initial and final states are the same. Among such processes are harmonic generations (both second and third), sum- and difference-frequency generations, four-wave mixing, etc. Nevertheless, we have not mentioned any examples of *nonparametric* nonlinear optical processes where the initial and final energy states are different. Among such processes, there are nonlinear absorption and some kinds of nonlinear scattering processes. Among the latter are *stimulated Raman* and *simulated Brillouin* scatterings. Covering the subject of nonlinear scattering is outside the scope of this book, and we refer the reader to nonlinear optics books for related reading [19]. However, nonlinear absorption processes deserve a few-word description in this book because these phenomena frequently accompany other nonlinear optical processes.

Nonlinear absorption can be attributed to different odd orders of nonlinear optical interactions, depending on the number of photons involved. As an example, Fig. 3.21 demonstrates two-photon absorption (2PA) [Part (a)] and three-photon absorption (3PA) [Part (b)]. Two-photon absorption is a third-order nonlinear optical process, while three-photon absorption is a fifth-order process. These processes entail simultaneous absorption of two and three photons, respectively, by optical nonlinearity, leading to an atom's excitation to a real excited state, shown with the solid lines in Fig. 3.21. In contrast, the intermediate states, shown in the figure with the dashed lines, are virtual. The initial and final states of 2PA and 3PA are different, signifying that these processes are nonparametric. Of course, it does not mean the violation of energy conservation law. It means that after the atoms' excitation, there is no optical dissipation that is part of the nonlinear optical process. The channels of energy dissipation after the excitation by multi-photon-absorption can be of optical or non-optical nature, but these are not part of the nonlinear optical process. As an example, the energy can be converted to heat, it can be spent on optical damage, or there can be free-carrier production (in the semiconductor material). As an

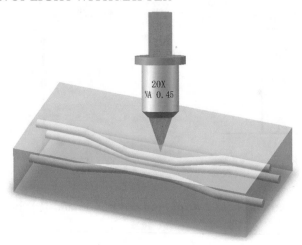

Figure 3.22: Femtosecond laser writing of optical waveguides in a bulk transparent material. The image is reproduced from Sugioka and Cheng, [30], with permission of AIP Publishing.

example of an optical dissipation channel, there can be fluorescence accompanying a nonlinear absorption process.

In general, multiphoton absorption processes are viewed as parasitic nonlinear optical processes that subtract useful power from the optical beams. On the other hand, these processes can be implemented in microfabrication. As one example, it is possible to permanently damage a piece of bulk material or optical fiber in a carefully controlled manner to obtain a region with modified optical properties that can control the flow of light. In such a way, optical waveguides and nanostructures can be created in silica glass and other materials by exposing them to powerful ultrashort laser pulses (see Fig. 3.22) [30]. In the figure, a powerful 800-nm red laser light is tightly focused by a microscopic objective to create microdamage in a piece of bulk material via multiphoton absorption. The defects are permanently written in the material, and they represent the regions of higher refractive index that can serve as channel waveguides guiding light.

This concludes our section on nonlinear optics where we attempted to summarize most notable nonlinear optical phenomena, and to provide a simple explanation to nonlinear optical effects. There are specialized books on nonlinear optics that one can explore for more advanced studying [19, 21, 22].

CHAPTER 4

Conclusions

This book covers aspects of light–matter interaction. In Chapter 1, we discussed the dual nature of light and various characteristics used to describe its color, strength, speed of propagation, frequency spectrum, and temporal behavior. Specifically, we looked at various physical manifestations of wave-particle duality, stressing that the wave nature of light appears in the situations where the light is bright. In contrast, its quantum (corpuscular) nature dominates in the experiments where very weak beams of light are used. We discussed a double-slit experiment where the photon counts are integrated over time. In the experiment, one starts with the proof of the corpuscular nature of light and ends up with the fact that the light exhibits wave nature in the same experiment, after letting the integration time be sufficiently long. This experiment served as a reconciliation of the two apparently contradictory ways of representing light. Both natures became evident under the same experimental conditions and were proved to co-exist simultaneously.

We next moved on to quantifying various properties of light, such as its color (using wavelength λ_0 and frequency ν), propagation speed (c_0 for free space and c for a medium), and wavenumber (k_0 in free space and k in a medium).

We discussed different ways of quantifying the strength of light. Specifically, we introduced the strength of the electromagnetic field associated with the light wave, expressed as the electric $\mathbf{E}(\mathbf{r}, t)$ and magnetic $\mathbf{H}(\mathbf{r}, t)$ fields. As these quantities are non-practical from the measurement standpoint (no devices can register optical fields in real-time), we introduced their measurable counterparts, such as the energy $\mathcal{E}$, optical power P, and optical intensity I. These quantities are proportional to the absolute value of the electric field, squared: $|\mathbf{E}|^2$. Existing detectors can readily measure the power and energy. Simultaneously, the intensity can be easily calculated using the value of the cross-section area of the optical beam (intensity = power per unit area).

We spoke about the wavefronts associated with a light wave, and how a light wave can be characterized based on what type of surfaces its wavefronts are. We mentioned plane wave approximation as the most convenient kind of wavefront to deal with (because all the light rays are parallel to each other, or collimated, and easier to control and quantify). We also discussed spherical waves and paraxial waves with more complicated wavefronts but with a very slow divergence of the light rays.

We made a distinction between monochromatic and broadband light. Monochromatic light has an infinite extent in time and is characterized by a single-frequency component. In contrast, broadband light could occupy an entire spectrum of frequencies. Monochromatic light is an idealistic concept and can only be used as an approximation. There are no optical waves with infinite duration, and their spectrum is never delta-function-like in the frequency domain. We discussed pulsed radiation where the light is emitted in the form of short temporal pulses and the relationship between the temporal pulse duration and the associated spectrum of light. We remarked that the shorter the temporal pulse duration is, the broader the corresponding spectrum of light.

In Chapter 2, we provided an overview of various optical materials at the level of their internal structures: crystalline or amorphous, molecular or atomic. Besides, we spoke about their conductive properties and their impact on the optical properties of materials. Specifically, we discriminated between three groups of materials: insulators with a wide bandgap, metals with essentially no bandgap and high rate of electron excitation and mobility as a consequence, and semiconductors with a relatively small bandgap as an intermediate case. Moreover, we introduced the three radiative atomic transitions: absorption, spontaneous, and stimulated emission, and explained the importance of the specific arrangement of energy levels associated with an optical material. We also spoke about various kinds of radiative transitions from the material standpoint: electronic and molecular rotational and vibrational transitions.

Chapter 3 was dedicated to an overview of light–matter interaction. We focused on the aspects of light–matter interaction without considering the specifics of optical components (such as interfaces between different media, engineered modifications to the flow of light, etc.). This topic is the subject matter of the second book in this series. Instead, we focused on the physics of the interaction of light with natural materials considered in bulk (non-structured). We started explaining how objects can appear in color as a consequence of their interaction with various frequencies of light. After that, we overviewed linear optical interaction of light with matter, limited to the effects of dispersion and absorption. We discussed the fact that absolutely all optical materials with no exception exhibit dispersion and possess resonances, and that the regions of high dispersion accompany the regions of high absorption in the spectrum of optical materials. We mentioned the importance of transparency windows of optical materials in relevance to passive optical components. We also explained that the material dispersion is the lowest within such spectral windows (between the material resonances). We discussed all the characteristics associated with optical media's material responses: their susceptibility χ, dielectric permittivity ϵ, refractive index n and absorption coefficient α. We identified real (n, α) and complex (χ, ϵ) parameters of the optical response, and presented the fact that the imaginary parts of χ and ϵ are associated with optical losses and are non-zero only in the frequency ranges where the losses are significant (in the vicinity of the material resonances).

After an extensive discussion of the linear optical regime of light–matter interaction, we covered some aspects of nonlinear optics where the parameters of optical response become intensity-dependent. We had a chance to observe the colorful effects of nonlinear optical interactions, associated with the changes in the frequency of light interacting with a nonlinear optical medium. Specifically, such effects include harmonic generation (second and third) and sum- and difference-frequency generation. We considered two lowest-order nonlinear optical contributions to the optical response of materials and limited our discussion to the effects arising from the second- and third-order nonlinear optical interactions. Apart from frequency mixing effects, we discussed self-action effects, such as self-focusing and self-phase modulation. We provided a brief discussion on parasitic and useful aspects of each nonlinear optical effect described in the chapter.

The future outlook includes a plan for several more books within this series. Book 2, *Introduction to Optics II: Passive Optical Components*, will cover the propagation of light through passive optical components, such as lenses, polarization optics, interference, and diffraction optics. Book 3, *Introduction to Optics III: Light-Emitting Devices*, will focus on active optical media, optical resonators and lasers, as well as optical amplifiers. The final book of the series, *Introduction to Optics IV: Optoelectronic Devices*, will cover optics of semiconductors, fundamentals of optoelectronics, and optoelectronic devices, such as semiconductor light-emitting devices and detectors. This topic will conclude our brief excursion to the fundamentals of optics and passive and active optical devices.

Bibliography

[1] S. Stierwalt, Einstein's legacy: The photoelectric effect, *Scientific American*, 2015. 3

[2] K. Mitchel, The theory of the surface photoelectric effect in metals-I, *Proc. of the Royal Society A*, 146, 1934. DOI: 10.1098/rspa.1934.0165. 3

[3] M. Ossiander, J. Reimensberger, S. Neppl, M. Mittermair, et al., Absolute timing of the photoelectric effect, *Nature*, 561:374–377, 2018. DOI: 10.1038/s41586-018-0503-6. 3

[4] R. H. Stuewer, Max planck, *Encyclopædia Britannica*, 2020. 3

[5] T. L. Dimitrova and A. Weis, The wave-particle duality of light: A demonstration experiment, *American Journal of Physics*, 76:137–142, 2008. DOI: 10.1119/1.2815364. 4

[6] Z. Zhang, M. Kushimoto, T. Sakai, N. Sugiyama, L. J. Schowalter, C. Sasaoka, and H. Amano, A 271.8 nm deep-ultraviolet laser diode for room temperature operation, *Applied Physics Express*, 12:124003, 2019. DOI: 10.7567/1882-0786/ab50e0. 6

[7] P. Chevalier, A. Amirzhan, F. Wang, M. Piccardo, S. G. Johnson, and F. Capasso, Widely tunable compact terahertz gas lasers, *Science*, 366:856–860 2019. DOI: 10.1126/science.aay8683. 6

[8] B. E. A. Saleh and M. C. Teich, *Fundamentals of Photonics*, 2nd ed., John Wiley & Sons, Inc., 2007. DOI: 10.1002/0471213748. 12, 13, 24, 25, 28, 53, 61

[9] D. Malacara, *Color Vision and Colorimetry: Theory and Applications*, 2nd ed., SPIE Press Book, 2011. DOI: 10.1117/3.881172. 20

[10] D. F. de Souza, P. P. F. da Silva, L. F. A. Fontenele, G. D. Barbosa, and M. de Olivera Jesus, Efficiency, quality, and environmental impacts: A comparative study of residential artificial lighting, *Energy Reports*, 5:409–424, 2019. DOI: 10.1016/j.egyr.2019.03.009. 22

[11] E. F. Schubert, J. Cho, and J. K. Kim, Light-emitting diodes, *Kirk–Othmer Encyclopedia of Chemical Technology*, 2015. DOI: 10.1002/0471238961.1209070811091908.a01.pub3. 22

[12] J. Cho, J. H. Park, J. K. Kim, and E. F. Schubert, White light-emitting diodes: History, progress and future, *Laser and Photonics Reviews*, 11:1600147, 2017. DOI: 10.1002/lpor.201600147. 22

[13] A. E. Siegman, *Lasers*, University Science Books, 1986. 22

[14] F. L. Pedrotti, L. M. Pedrotti, and L. S. Pedrotti, *Introduction to Optics*, 3rd ed., Cambridge University Press, 2018. DOI: 10.1017/9781108552493. 28

[15] E. Hecht, *Optics*, 5th ed., Pearson, 2017. 28

[16] R. L. Brooks, *The Fundamentals of Atomic and Molecular Physics*, Springer, 2013. DOI: 10.1007/978-1-4614-6678-9. 39

[17] W. Demtröder, *Atoms, Molecules and Photons*, 3rd ed., Springer, 2018. DOI: 10.1007/978-3-662-55523-1. 39

[18] B. Tatian, Fitting refractive-index data with the Sellmeier dispersion formula, *Applied Optics*, 23:4477–4485, 1984. DOI: 10.1364/ao.23.004477. 56

[19] R. W. Boyd, *Nonlinear Optics*, 4th ed., Academic Press, 2020. DOI: 10.1016/C2015-0-05510-1. 61, 65, 76, 83, 84

[20] K. Dolgaleva, N. Lepeshkin, and R. W. Boyd, *Frequency Doubling*, Routlidge, 2006. 61, 62

[21] Y. R. Shen, *The Principles of Nonlinear Optics*, 3rd ed., Wiley, 2003. 65, 84

[22] N. Bloembergen, *Nonlinear Optics*, 4th ed., World Scientific, 1996. DOI: 10.1142/3046. 65, 84

[23] W. Zhang, H. Yu, H. Wu, and P. S. Halasyamani, Phase-matching in nonlinear optical compounds: A materials perspective, *Chemistry of Materials*, 29:2655–2668, 2017. DOI: 10.1021/acs.chemmater.7b00243. 72

[24] C. Monat, M. de Sterke, and B. J. Eggleton, Slow light enhanced nonlinear optics in periodic structures, *Journal of Optics*, 12:104003, 2010. DOI: 10.1088/2040-8978/12/10/104003. 72

[25] J. Bravo-Abad, A. Rodriguez, P. Bermel, S. G. Johnson, and J. D. Joannopoulos, Ehnanced nonlinear optics in photonic-crystal microcavities, *Optics Express*, 15:16161–16176, 2007. DOI: 10.1364/oe.15.016161. 72

[26] M. Kauranen and A. V. Zayats, Nonlinear plasmonics, *Nature Photonics*, 6:737–748, 2012. DOI: 10.1038/nphoton.2012.244. 72

[27] Z. Lin, X. Jiang, L. Kang, P. Gong, and S. Luo, First-principles materials applications and design of nonlinear optical crystals, *Journal of Physics D: Applied Physics*, 47:253001, 2014. DOI: 10.1088/0022-3727/47/25/253001. 74

[28] Y. Zhang, Y. Leng, J. Liu, N. Ji, X. Duan, J. Li, X. Zhao, J. Wang, and H. Jiang, Mechanism of hydrogen treatment in $KTiOPO_4$ crystals at high temperature: Experimental and first-principles studies, *CrystEngComm*, 17:3793–3799, 2015. DOI: 10.1039/c5ce00071h. 74

[29] T. J. Kippenberg, R. Holzwarth, and S. A. Diddams, Frequency comb generation via cascaded FWM processes, *Science*, 29:555–559, 2011. 82

[30] K. Sugioka and Y. Cheng, Femtosecond laser three-dimensional micro- and nanofabrication, *Applied Physics Reviews*, 1:041303, 2014. DOI: 10.1063/1.4904320. 84

Author's Biography

KSENIA DOLGALEVA

Ksenia Dolgaleva completed her undergraduate program and Diploma in Physics (an equivalent of a Master's degree) at the Faculty of Physics at Lomonosov Moscow State University. Her field of undergraduate specialization was laser physics. She performed her doctoral studies at the Institute of Optics, the University of Rochester, in the research group of Prof. Robert Boyd. Since then, nonlinear optics has become her primary subject of interest: all her research is centered on this field.

After completing her Ph.D. in Optics at the University of Rochester, Ksenia performed her postdoctoral research with Prof. Stewart Aitchison at the Department of Electrical and Computer Engineering, the University of Toronto. She worked on integrated photonic devices based on semiconductor materials.

Dr. Dolgaleva is presently an Associate Professor at the University of Ottawa and a Tier 2 Canada Research Chair in Integrated Photonics, and a recipient of the Ontario Early Researchers award. Her primary affiliation is with the School of Electrical Engineering and Computer Science, Faculty of Engineering. She is also cross-appointed with the Department of Physics, Faculty of Science. Her current research fields include nonlinear photonics in III-V semiconductors, the interaction of light with matter, terahertz optics, and nanostructured optical materials. She has expertise in a variety of optics subfields, which she is passionate to share with readers in this book series.

Printed in the United States
by Baker & Taylor Publisher Services